Properties of Materials
for Electrical Engineers

Properties of Materials for Electrical Engineers

K. J. PASCOE

*University Lecturer in Engineering
and Fellow of St. John's College,
Cambridge*

JOHN WILEY & SONS

Chichester · New York · Brisbane · Toronto

Copyright © 1973 John Wiley & Sons Ltd. All rights reserved. No part of this publication may be reproduced, stored in a retrieval system, or transmitted, in any form or by any means, electronic, mechanical photocopying, recording or otherwise, without the prior written permission of the Copyright owner.

Library of Congress Catalog card No. 72-8612

ISBN 0 471 66910 5 Cloth bound

ISBN 0 471 66911 3 Paper bound

REPRINTED 1974
REPRINTED 1978
REPRINTED 1980

Set on Monophoto Filmsetter and printed by J. W. Arrowsmith Ltd., Bristol, England.

Preface

While all engineers are, in their work, dependent upon and limited by the physical properties of available materials, the electrical engineer has, perhaps, the widest range of properties with which to be concerned. Proper application and efficient use of materials can be achieved successfully only by having an understanding of the factors that determine the properties, which are linked with the microscopic and sub-microscopic structure of the material—down to the structure of the atom itself.

The purpose of this book is to introduce the reader to this structure of materials and to elaborate particular facets to the stage where he can understand the function and behaviour of electrical materials and the ever growing number of ways in which they are applied.

A treatment is given of the fundamental nature of elementary particles—their dual personality of behaving both as waves and particles, upon which an understanding of solid-state physics depends. This leads to the formation of atoms from the elementary particles and the bonding between atoms, after which the behaviour of atoms in large numbers in the gaseous, liquid and solid states are considered. The later chapters give, in terms of solid-state science, explanations of the electrical and magnetic properties and behaviour of materials and discusses their applications.

No attempt has been made to include all solid-state devices nor to give details of the method of using them. The reader must refer to more specialized texts for such information.

SI units have, in general, been used throughout. The few exceptions are other units which still have wide use, such as ångström and electron-volt. Problems are given at the end of most chapters.

Cambridge K. J. PASCOE
August 1972

Acknowledgements

The publishers acknowledge permission for the use of the following figures and tables which are based on diagrams or data in the works detailed:

Figure 10.14,   A. B. Pippard, Phil. Trans. Roy. Soc.
Figure 12.14,   H. D. Megaw, *Ferroelectricity in Crystals*, Methuen.
Figures 14.13,   14.14, 14.16, 14.23, 14.24, Characteristics of OA 10, OAZ 203, ORP 90 and VA 1011, Mullard Ltd.

Table 14.1,   H. F. Wolf, *Semiconductors*, Wiley-Interscience, 1971, and other sources.
Table 15.1,   J. E. Goldman, Ed., *The Science of Engineering Materials*, Wiley, 1957.

The author gratefully acknowledges permission given by the Syndics of the Cambridge University Press to reproduce certain questions from Examination papers.*

The sources of these questions are indicated as follows:

E   Engineering Tripos
EST   Electrical Sciences Tripos
MST   Mechanical Science Tripos
P   Preliminary Examination in Mechanical Sciences

---

* Complete sets of examination papers are published by Cambridge University Press, Bentley House, 200 Euston Road, London NW1 2DB.

Contents

Chapter One: The Chemical Elements
1.1 Introduction 1
1.2 The periodic table 1
1.3 Valency 6
1.4 Atomic number 6

Chapter Two: Wave-particle Duality
2.1 Introduction 9
2.2 Black-body radiation 9
✗ 2.3 Photoelectric effect 11
✗ 2.4 Bohr–Rutherford atom 12
2.5 Compton effect 15
2.6 de Broglie's hypothesis 18
2.7 Electron diffraction 18
2.8 Representation of a photon or particle by a wave . . . 18
2.9 Heisenberg's uncertainty principle 21
2.10 Simple wave equations 24
2.11 Wave representation of a beam of electrons 26
2.12 Formulation of Schrödinger's equation 28
2.13 Conditions for solution of Schrödinger's equation . . . 29
2.14 Normalization 30

Chapter Three: Wave Mechanics Applied to Single Electrons
3.1 Motion of a beam of electrons through potential jumps . . 32
3.2 General consideration of stationary states 38
3.3 Stationary states in a potential well 40
3.4 Three-dimensional potential well 45
3.5 Particle moving in a central field 47
3.6 Simple harmonic linear oscillator 52

Chapter Four: The Many-electron Atom
- 4.1 Wave equation for two or more particles 56
- 4.2 The determination of energy levels 56
- 4.3 Pauli's exclusion principle and electron spin . . . 57
- 4.4 Splitting of energy levels 58
- 4.5 Electron structure of the elements 59
- 4.6 X-ray spectra 60

Chapter Five: Interatomic Bonding
- 5.1 Overlapping wave functions 63
- 5.2 Hydrogen molecule–covalent bonding 64
- 5.3 Force between inert gas molecules 68
- 5.4 Ionic or polar bonding 69
- 5.5 Covalent and ionic bonds 70
- 5.6 Directional properties of covalent bonds 71
- 5.7 Metallic bonding 76
- 5.8 Intermolecular forces 76

Chapter Six: Aggregations of Atoms—Gases and Liquids
- 6.1 Introduction 79
- 6.2 The behaviour of a gas 79
- 6.3 The kinetic theory of gases 80
- 6.4 Mole 81
- 6.5 Maxwellian distribution of velocities 81
- 6.6 Mean free path 82
- 6.7 Transport properties 84
- 6.8 Variation of λ, μ and K with pressure 88
- 6.9 Deviations from the gas laws 88
- 6.10 Van der Waals' equation 89
- 6.11 Comparison of Van der Waals' equation with experiment . 90
- 6.12 The behaviour of liquids 90
- 6.13 Surface tension 91
- 6.14 Vapour pressure 92
- 6.15 Viscosity 92
- 6.16 Thermal conductivity 93

Chapter Seven: The Crystalline State
- 7.1 Introduction 95
- 7.2 Space lattices 95
- 7.3 Indices of planes 98
- 7.4 Indices of direction 100
- 7.5 Crystal systems 101

Contents ix

7.6 Face-centred cubic structure 102
7.7 Body-centred cubic structure 104
7.8 Close-packed hexagonal structure 106
7.9 The diamond structure 107
7.10 Ionic crystals 108
7.11 Relation of crystal structure to properties 108

Chapter Eight: Solid Compounds
8.1 Introduction 111
8.2 Ceramics 111
8.3 Physical properties of ceramics 112
8.4 Vitreous structures 115
8.5 Glass-ceramics 117
8.6 Organic compounds 117
8.7 Saturated hydrocarbons 117
8.8 Unsaturated hydrocarbons 119
8.9 Aromatic hydrocarbons 120
8.10 Polymerization 121
8.11 Some common polymers 122
8.12 Relationship of mechanical properties to polymer shape . 124
8.13 Elastomers 125
8.14 Epoxy resins 126
8.15 Fillers 127
8.16 Plasticizers 128

Chapter Nine: Thermodynamics of Materials
9.1 Introduction 129
9.2 Kinetic energy of a gas molecule 129
9.3 Specific heat of a gas 131
9.4 Specific heat of a solid 131
9.5 Thermal expansion 133
9.6 Thermal equilibrium 134
9.7 Stable and metastable states 135
9.8 Activation energy 135
9.9 Diffusion 137

Chapter Ten: Band Structure of Solids
10.1 Introduction 140
10.2 Energy bands 140
10.3 The band structure of the alkali metals 141
10.4 The band structure of the alkaline earth metals . . 142
10.5 Occupancy of energy levels 143

10.6	Distribution of available energy states	144
10.7	The effect of the periodic field	146
10.8	The three-dimensional periodic field	149
10.9	The Fermi surface	153
10.10	The relation of energy bands and conductivity	155

Chapter Eleven: Conducting Materials

11.1	Electrical conductivity	158
11.2	Mobility and resistivity	159
11.3	Factors affecting resistivity	161
11.4	Thermal conductivity	163
11.5	Work function	164
11.6	Thermionic emission	165
11.7	Schottky effect	166
11.8	Thermoelectricity	167

Chapter Twelve: Insulating Materials

12.1	Dielectrics	173
12.2	Dielectric constant	174
12.3	Polarizability	176
12.4	Temperature dependence	178
12.5	Frequency response of polarization	180
12.6	Conductivity of dielectrics	183
12.7	Electrical breakdown	184
12.8	Insulating materials	184
12.9	Piezoelectricity	186
12.10	Electrostriction	188
12.11	Pyroelectricity	188
12.12	Ferroelectricity	189
12.13	Effect of temperature	191
12.14	Ferroelectric materials	192
12.15	Barium titanate	192

Chapter Thirteen: Semiconductors

13.1	The energy bands of diamond	195
13.2	Group IV semiconductors	196
13.3	Density of states in a semiconductor	197
13.4	Effect of impurity atoms in a semiconductor	200
13.5	Position of the Fermi level	205
13.6	Conductivity in semiconductors	207
13.7	Hall effect	209
13.8	Diffusion	212

Contents

Chapter Fourteen: Semiconductor Materials and Devices
- 14.1 Semiconductor materials 215
- 14.2 Purification of semiconductor materials 218
- 14.3 Minority carrier lifetime 219
- 14.4 Surface states 220
- 14.5 Contacts 222
- 14.6 Rectifying contacts 222
- 14.7 Ohmic contacts 225
- 14.8 The p–n junction 226
- 14.9 Applications of p–n junctions 229
- 14.10 The junction transistor 234
- 14.11 The field-effect transistor (FET) 235
- 14.12 The metal-oxide semiconductor transistor (MOST) . . 236
- 14.13 Other semiconductor applications 237
- 14.14 Microminiature solid state circuits 240

Chapter Fifteen: Magnetic Materials
- 15.1 Basic concepts 246
- 15.2 Atomic magnets 248
- 15.3 Diamagnetism 250
- 15.4 Paramagnetism 252
- 15.5 Paramagnetism of simple metals 253
- 15.6 Ferromagnetism 254
- 15.7 Exchange forces 255
- 15.8 Anisotropy energy 257
- 15.9 Magnetostriction 258
- 15.10 Domains 259
- 15.11 Ferromagnetic materials 261
- 15.12 Application of ferromagnetic materials 263
- 15.13 Exchange mechanisms 265
- 15.14 Susceptibility of antiferromagnetic material . . . 266
- 15.15 Ferrimagnetism 267
- 15.16 Ferrimagnetic materials 268
- 15.17 Spin resonance 271
- 15.18 Electromagnetic wave propagation in ferrites . . . 271
- 15.19 Applications of ferrites 273

Appendix 1 SI units 275
Appendix 2 Fundamental constants and conversion constants . . 277
Appendix 3 Solution of Schrödinger's equation for the hydrogen atom 278
Appendix 4 Solution of Schrödinger's equation for simple harmonic oscillator 285
Appendix 5 Proof of spherical symmetry of completely full p and d levels 288

Appendix 6   The electron structure of the elements . . . 290
Appendix 7   The Maxwell–Boltzmann distribution . . . 293
Appendix 8   Maxwell–Boltzmann and Fermi–Dirac statistics . . 295
Appendix 9   Proof of the Richardson–Dushmann equation. . . 298
Appendix 10 Wave-mechanical treatment of orbital and spin motions . 302
Appendix 11 The Brillouin function 305
Appendix 12 The Curie–Weiss law 307
Appendix 13 The origin of ferromagnetism 308
Answers to numerical questions 311
Bibliography 314
Index 317

CHAPTER ONE

The Chemical Elements

1.1 Introduction

A chemical element is defined as a substance which cannot be broken down into simpler substances by ordinary means and chemical elements are distinguished from each other by differences in the manner in which they react chemically. A chemical compound is a substance which is a combination of two or more elements.

Ninety-two distinct elements have been found to occur in nature and several more have been made artificially.

The smallest unit of an element that still has the properties of that element is an *atom*. Similarly, the smallest unit of a compound that still has the properties of that compound is a *molecule*. The structure of an atom is discussed in Sections 1.4 and 2.4 and the formation of molecules from atoms in Chapter 5. The mass of an atom in terms of the unified atomic mass unit [see Appendix 2 (p. 277)] is referred to as the *atomic weight* and the mass of a molecule in terms of the same unit as the *molecular weight*. This unit has been chosen as one twelfth of the mass of the carbon atom having mass number 12 (see Section 2.4).

The atoms of a particular element will all have the same atomic weight or one of a definite set of values. Atoms of the same element with different atomic weights are known as *isotopes*. In nature, the proportions of the various isotopes of an element are usually constant and the atomic weight as determined from gravimetric experiments is a mean value. The different isotopes of an element can be separated by physical, but not usually by chemical, means.

A list of all the named elements arranged in alphabetical order is given in Table 1.1, together with their chemical symbols, atomic weights and numbers, melting and boiling points and data concerning their crystal structure.

1.2 The periodic table

If the elements are listed in order of increasing atomic weight, it is found that elements with similar chemical properties occur at regular intervals. Mendeleev

Table 1.1 Some physical properties of the elements

Element	Symbol	Atomic number	Atomic weight[a]	M.P. °C[b]	B.P. °C[b]	Crystal structure[c]	Lattice constants[d] a	b	c
Actinium	Ac	89	[227]	1050		f.c.c.	5·311		
Aluminium	Al	13	26·9815	660	2400	f.c.c.	4·0496		
Americium	Am	95	[243]	850					
Antimony	Sb	51	121·75	630	1440	r.	4·5076 α = 57° 6·5′		
Argon	A	18	39·948	−189	−186	f.c.c.	5·42 (at −233°C)		
Arsenic	As	33	74·9216	814[e]		r.	4·139 α = 54° 7·5′		
Astatine	At	85	[210]						
Barium	Ba	56	137·34	710	1770	b.c.c.	5·025		
Berkelium	Bk	97	[247]						
Beryllium	Be	4	9·0122	1280	2450	c.p.h.	2·2856		3·5843
Bismuth	Bi	83	208·980	271	1530	r.	4·7450 α = 57° 14·2′		
Boron	B	5	10·811	2300	2550	t.	8·73		5·03
Bromine	Br	35	79·909	−7	58	orth.	4·48	6·67	8·72
							(at −150°C)		
Cadmium	Cd	48	112·40	321	767	c.p.h.	2·9793		5·618
Caesium	Cs	55	132·905	29	713	b.c.c.	6·16		
Calcium	Ca	20	40·08	850	1440	f.c.c.	5·582		
Californium	Cf	98	[249]						
Carbon (graphite)	C	6	12·01115	3500	3900	hex.	2·4612		6·7079
Cerium	Ce	58	140·12	804	2900	f.c.c.	5·1615		
Chlorine	Cl	17	35·453	−101	−34	t.	8·56 (at −185°C)		6·12
Chromium	Cr	24	51·996	1900	2600	b.c.c.	2·8850		
Cobalt	Co	27	58·9322	1492	2900	c.p.h.	2·5053		4·0886
Copper	Cu	29	63·54	1083	2550	f.c.c.	3·6150		
Curium	Cm	96	[247]						
Dysprosium	Dy	66	162·50	1500	2600	c.p.h.	3·5903		5·6475
Einsteinium	Es	99	[254]						
Erbium	Er	68	167·26	1525	2600	c.p.h.	3·5588		5·5874
Europium	Eu	63	151·96	900	1400	b.c.c.	4·606		
Fermium	Fm	100	[253]						
Fluorine	F	9	18·9984	−220	−188				
Francium	Fr	87	[223]						
Gadolinium	Gd	64	157·25	1320	2700	c.p.h.	3·6360		5·7826
Gallium	Ga	31	69·72	30	2250	orth.	4·524	4·523	7·661
Germanium	Ge	32	72·59	958	2880	d.	5·6575		
Gold	Au	79	196·967	1063	2660	f.c.c.	4·0786		
Hafnium	Hf	72	178·49	2000	5100	c.p.h.	3·1969		5·0583
Helium	He	2	4·0026	−270	−269	c.p.h.?	3·57		5·83
							(at −271·5°C)		
Holmium	Ho	67	164·930	1500	2700	c.p.h.	3·5773		5·6158
Hydrogen	H	1	1·00797	−259	−253	hex.	3·75 (at −271°C)		6·12
Indium	In	49	114·82	156	2000	f.c.t.	3·2515		4·9459
Iodine	I	53	126·9044	114	183	orth.	4·792	7·271	9·773
Iridium	Ir	77	192·2	2443	5300	f.c.c.	3·8394		
Iron	Fe	26	55·847	1539	2900	b.c.c.	2·8663		
Krypton	Kr	36	83·80	−157	−153	f.c.c.	5·68 (at −191°C)		
Lanthanum	La	57	138·91	920	4200	c.p.h.	3·761		6·061
Lawrencium	Lw	103	[257]						

Table 1.1 (cont.)

Element	Symbol	Atomic number	Atomic weight[a]	M.P. °C[b]	B.P. °C[b]	Crystal structure[c]	Lattice constants[d]		
							a	b	c
Lead	Pb	82	207·19	327	1750	f.c.c.	4·9505		
Lithium	Li	3	6·939	180	1330	b.c.c.	3·5089		
Lutetium	Lu	71	174·97	1700	1900	c.p.h.	3·5031		5·5509
Magnesium	Mg	12	24·312	650	1100	c.p.h.	3·2094		5·2103
Manganese	Mn	25	54·938	1250	2100	cub.	8·912		
Mendelevium	Mv	101	[256]						
Mercury	Hg	80	200·59	−39	357	r.	3·005 $\alpha = 70° 31·7'$ (at −46 °C)		
Molybdenum	Mo	42	95·94	2620	4600	b.c.c.	3·1468		
Neodymium	Nd	60	144·24	1024	3170	c.p.h.	3·6579		5·899
Neon	Ne	10	20·183	−249	−246	f.c.c.	4·52 (at −268 °C)		
Neptunium	Np	93	[237]	640					
Nickel	Ni	28	58·71	1453	2820	f.c.c.	3·5241		
Niobium (Columbium)	Nb (Cb)	41	92·906	2420	5100	b.c.c.	3·3007		
Nitrogen	N	7	14·0067	−210	−196	hex.	4·03 (at −234 °C)		6·59
Nobelium	No	102	[253]						
Osmium	Os	76	190·2	2700	4600	c.p.h.	2·7314		4·3197
Oxygen	O	8	15·9994	−219	−183	cub.	6·83 (at −225 °C)		
Palladium	Pd	46	106·4	1552	3200	f.c.c.	3·8898		
Phosphorus	P	15	30·9738	44	280	cub.	7·17 (at −35 °C)		
Platinum	Pt	78	195·09	1769	3800	f.c.c.	3·9231		
Plutonium	Pu	94	[242]						
Polonium	Po	84	[210]	254	960	r.	3·36 $\alpha = 98° 13'$		
Potassium	K	19	39·102	63	760	b.c.c.	5·333		
Praseodymium	Pr	59	140·907	935	3000	c.p.h.	3·6725		5·917
Promethium	Pm	61	[147]						
Protactinium	Pa	91	[231]	3000					
Radium	Ra	88	226·05	700	1140				
Radon	Rn	86	[222]	−71	−62				
Rhenium	Re	75	186·2	3170	5900	c.p.h.	2·760		4·458
Rhodium	Rh	45	102·905	1960	3900	f.c.c.	3·8031		
Rubidium	Rb	37	85·47	39	710	b.c.c.	5·70		
Ruthenium	Ru	44	101·07	2400	3900	c.p.h.	2·7058		4·2819
Samarium	Sm	62	150·35	1052	1600				
Scandium	Sc	21	44·956	1400	2500	f.c.c.	4·533		
Selenium	Se	34	78·96	217	685	hex.	4·3640		4·9594
Silicon	Si	14	28·086	1410	2480	d.	5·4305		
Silver	Ag	47	107·870	961	2180	f.c.c.	4·0862		
Sodium	Na	11	22·9898	98	883	b.c.c.	4·2906		
Strontium	Sr	38	87·62	770	1460	f.c.c.	6·075		
Sulphur	S	16	32·064	119	445	f.c.orth.	10·437	12·845	24·369
Tantalum	Ta	73	180·948	3000	6000	b.c.c.	3·3058		
Technetium (Masurium)	Tc (Ma)	43	[99]	2700					
Tellurium	Te	52	127·60	450	997	hex.	4·4565		5·9268
Terbium	Tb	65	158·924	1450	2500	c.p.h.	3·6010		5·6936
Thallium	Tl	81	204·37	304	1460	c.p.h.	3·4560		5·5248
Thorium	Th	90	232·038	1700	4200	f.c.c.	5·0843		

Table 1.1 (cont.)

Element	Symbol	Atomic number	Atomic weight[a]	M.P. °C[b]	B.P. °C[b]	Crystal structure[c]	Lattice constants[d] a	b	c
Thulium	Tm	69	168.934	1600	2100	c.p.h.	3.5375		5.5546
Tin	Sn	50	118.69	232	2606	b.c.t.	5.8313		3.1812
Titanium	Ti	22	47.90	1680	3300	c.p.h.	2.9504		4.6833
Tungsten	W	74	183.85	3380	5700	b.c.c.	3.1652		
Uranium	U	92	238.03	1133	3800	orth.	2.858	5.877	4.945
Vanadium	V	23	50.942	1920	3400	b.c.c.	3.0282		
Xenon	Xe	54	131.30	−112	−108	f.c.c.	6.24 (at −185°C)		
Ytterbium	Yb	70	173.04	824	1500	f.c.c.	5.486		
Yttrium	Y	39	88.905	1500	3000	c.p.h.	3.6474		5.7306
Zinc	Zn	30	65.37	419	907	c.p.h.	2.6649		4.9468
Zirconium	Zr	40	91.22	1850	4400	c.p.h.	3.2312		5.1476

[a] The values of atomic weights are those recommended in the Report of the International Commission on Atomic Weights (1961). These are for the naturally occurring mixture of isotopes, except where a value is given in brackets. Each of these denotes the mass number of the isotope of longest known half-life, which is not necessarily the most important in atomic energy work. The atomic weights are in terms of the unified atomic mass unit (see Appendix 2). Because of the natural variation in the relative abundance of their isotopes, the atomic weights of boron and sulphur have ranges of ±0.003 about the values quoted.

[b] Melting and boiling points are given to the nearest degree. Except for the noble metals, most values above about 1500°C are not known accurately.

[c] The structure given is that at room temperature, except for elements which are not solid at that temperature. Many elements have other structures at higher temperature. Structures are denoted as follows:

    b.c.c.   body-centred cubic,
    b.c.t.   body-centred tetragonal,
    c.p.h.   close-packed hexagonal,
    cub.   cubic,
    d.   diamond structure, two interpenetrating f.c.c. lattices,
    f.c.c.   face-centred cubic,
    f.c.orth.   face-centred orthorhombic,
    f.c.t.   face-centred tetragonal,
    hex.   hexagonal,
    monoc.   monoclinic,
    orth.   orthorhombic,
    r.   rhombohedral,
    t.   tetragonal

[d] Lattice constants are given for 20°C unless otherwise stated. The values are in ångströms (1 ångström, Å = 10^{-10} m).

[e] Value at 36 atmospheres pressure. Arsenic sublimes at 610°C at a pressure of one atmosphere.

published, in 1869, his *law of periodicity* that the properties of the elements, as well as the properties of their compounds, form periodic functions of the atomic weight of the elements. From this follows the *periodic table* in which the elements with similar properties are arranged in groups. At that time, many of the naturally-occurring elements had not been discovered and gaps had to be left in the table among the higher atomic weight elements to preserve the grouping. The periodic table including all the naturally occurring elements is shown in Table 1.2.

Table 1.2 The periodic table of the elements (after Mendeleev)

Period	Group I		Group II		Group III		Group IV		Group V		Group VI		Group VII		Group VIII	Group 0
	A	B	A	B	A	B	A	B	A	B	A	B	A	B		
1	H															He
2	Li		Be		B		C		N		O		F			Ne
3	Na		Mg		Al		Si		P		S		Cl			A
4	K	Cu	Ca	Zn	Sc	Ga	Ti	Ge	V	As	Cr	Se	Mn	Br	Fe Co Ni	Kr
5	Rb	Ag	Sr	Cd	Y	In	Zr	Sn	Nb	Sb	Mo	Te	Ma	I	Ru Rh Pd	Xe
6	Cs	Au	Ba	Hg	La and rare earths	Tl	Hf	Pb	Ta	Bi	W	Po	Re	—	Os Ir Pt	Rn
7	—		Ra		Ac		Th		Pa		U					

There are nine groups each containing elements which show similarity of properties. The elements are listed in order of increasing atomic weight to form seven periods. The first period contains only two elements—hydrogen and helium. The second and third periods contain eight elements each, the fourth and fifth periods contain eighteen elements each, there are thirty-two in the sixth period and the seventh is incomplete. The last element in each period, i.e. in Group 0, is a chemically inert gas. The remaining elements show a gradual transition from the strongly electropositive alkali elements in Group I (lithium, sodium, etc.) to the strongly electronegative halogens in Group VII (fluorine, chlorine, etc.).

The elements of each group in the fourth and subsequent periods are divided into two sub-groups, the elements of one sub-group more closely resembling the elements of that group in the second and third periods than do those of the other sub-group. For example in Group I, potassium, rubidium and caesium have a greater resemblance to lithium and sodium than do copper, silver and gold. The elements in Group VIII have properties which are intermediate between those of Group VIIA and Group IB elements. Thus, iron, cobalt and nickel show a gradual transition of properties between those of manganese and copper.

Many physical properties are also periodic functions of the atomic number (see Section 1.4). Atomic volume (atomic weight/specific gravity) is one of these, having exceptionally high values for the alkali metals. Others are melting and boiling points, specific gravity, hardness, thermal and electrical conductivities. Specific heat, which is considered in Sections 9.3 and 9.4, is not a periodic property.

1.3 Valency

The valency of an element is the number of atoms of hydrogen with which one atom of it will combine or which one atom will replace.

Valencies, in general, rise from zero in Group 0 to four in Group IV and either fall again to one or increase to seven in Group VII. Thus elements of Groups V, VI and VII may exhibit two valencies which total to eight. Table 1.3 shows examples of compounds formed by representative elements in each group which illustrate these values of valency.

Table 1.3 Valencies of representative elements

Group	0	I	II	III	IV	V	VI	VII
Valency	0	1	2	3	4	5 3	6 2	7 1
Element	He	Li	Be	B	C	N	S	Cl
Compound	—	LiH Li_2O	BeO	B_2O_3	CH_4 CO_2	NH_3 N_2O_5	H_2S SO_2	HCl Cl_2O_7

In addition, each of the transition elements (see Section 1.4) may show several valencies.

The similarities and dissimilarities of the elements must depend upon differences in the atoms of the various elements and so an explanation is dependent upon a knowledge of the structure of the atom.

1.4 Atomic number

Each atom consists of a central positively-charged nucleus around which circulate a number of electrons, sufficient to give electrical neutrality to the atom. The nuclear charge is $+Ze$, where $-e$ is the charge on an electron and Z is an integer. Z is thus equal to the number of orbiting electrons and is known as the *atomic number*, each element having a different atomic number.

In general, the atomic weight increases as the atomic number increases. There are a few exceptions, where the order of elements had to be reversed from

The Chemical Elements

that given by their atomic weights to make the properties fit into the periodic table. In Table 1.2 the elements are strictly in order of atomic number.

A more modern layout for the periodic table is shown in Figure 1.1. The

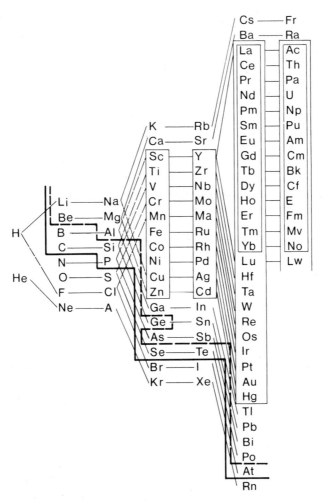

Figure 1.1. Periodic table. The bold dashed line separates metals from metalloids; the bold full line separates metalloids from non-metals

elements in each period are arranged in a single vertical line and elements in different periods with similar chemical and physical properties are linked by cross-lines.

The first period, comprising hydrogen and helium only, is followed by two periods of eight elements each. Each element in the third period falls alongside an element of the second period with very similar properties. Potassium and calcium of the fourth period bear strong resemblances to sodium and magnesium, respectively, and are followed by ten more elements before gallium, which resembles aluminium. The following five elements correspond to the last five elements of the third period. Also as was stated in Section 1.2, sodium shows a weaker resemblance to copper, magnesium to zinc, etc. These weaker resemblances are shown by broken lines in Figure 1.1. The ten elements, scandium to zinc, are known as *transition elements*. Almost all of these have important and widespread engineering applications.

The fifth period is also a long period with all elements in it resembling the respective ones of the fourth period. The sixth period commences in a manner similar to the fourth and fifth, but at the beginning of the transition elements there appear fifteen *rare earth* elements, from lanthanum to lutetium. These resemble yttrium and each other so closely that they are placed in a single space of the periodic table shown in Table 1.2. After these rare earths or *lanthanides*, the sixth period follows the previous one. The seventh period is incomplete but appears to follow the pattern of the previous one. The elements following radium, which are similar to each other and analogous to the rare earths, are known as the *actinides*. Uranium is the highest atomic number element found in nature. Thirteen trans-uranium elements with atomic numbers from 93 to 105 have been produced artificially by nuclear reactions and eleven of them are shown in Figure 1.1.

It should be observed that the numbers in the periods are 2, 8, 8, 18, 18, 32, —, which are 2×1^2, 2×2^2 twice, 2×3^2 twice and 2×4^2.

An alternative classification of elements is on a basis of whether they show metallic, non-metallic or intermediate metalloid properties. Apart from the trans-uranium elements the numbers in the three groups are 70, 14 and 8, respectively. The picture of atomic structure developed later in the book will explain not only the cause of metallic and non-metallic properties but also why there are so many metals.

Before this picture can be developed, however, it is necessary to study the nature of the fundamental components of atoms and the manners in which they interchange energy. This takes us into the realm of modern physics and the alternative wave and corpuscular natures which are exhibited both by the fundamental components and by radiant energy. The development of this duality concept and its application are considered in the next chapter.

Chapter Two

Wave-particle Duality

2.1 Introduction

At the beginning of the twentieth century, a sufficient and satisfactory explanation of most of the observed phenomena of physics could be given by Newton's laws of mechanics and Maxwell's laws relating to electromagnetic waves. As more was discovered about atomic structure and the interaction of electromagnetic waves with matter, these laws were no longer able to give a complete explanation of the observed facts. It was not merely a matter of minor discrepancies, but there were fundamental differences between observation and prediction which were to be explained later by the concept of a dual nature of particles and waves. The development of this concept, which is the basic idea of wave mechanics, form the subject matter of this chapter.

2.2 Black-body radiation

Every body emits radiation, which increases in intensity as its temperature is increased. It also absorbs radiation from its surroundings so that the net effect is not significant until the temperature of the body is well above the ambient temperature.

The maximum intensity of radiation that could be emitted by a body at any temperature is known as *black-body radiation*. This would be emitted by a perfect radiant surface and is the same as that emitted from the inside of a hollow body through a small aperture in its wall. Generally, the radiation emitted from a surface is dependent on the material and the nature of the surface, but black-body radiation is independent of these.

The rate of emission of energy per unit area in black-body radiation increases with increasing temperature and the distribution of the energy among different wavelengths varies with temperature. The spectral distribution curves for some temperatures are shown in Figure 2.1. A study of these experimentally determined curves reveals two facts. The first, known as *Wien's displacement law* is that the wavelength corresponding to the maximum of a curve varies in-

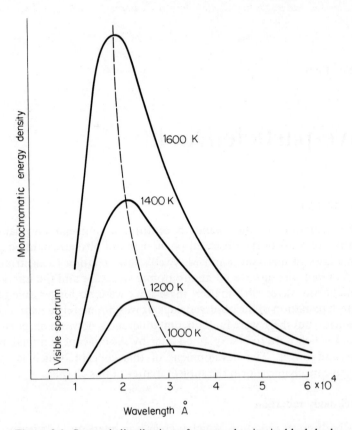

Figure 2.1. Spectral distribution of energy density in black-body radiation at various temperatures.

versely with temperature, i.e. $\lambda_{max} \propto 1/T$ and the curves are of similar form so that corresponding points on different curves are related by

$$\lambda_1 T_1 = \lambda_2 T_2$$

The second is that the rate of emission of the total energy E from unit area of a black body, which is the area under the appropriate curve in Figure 2.1, is found to vary as the fourth power of the absolute temperature T and is expressed in the *Stefan's law*

$$E = \sigma T^4$$

The Stefan–Boltzmann constant σ has the value $5 \cdot 67 \times 10^{-8}$ J m^{-2} K^{-4} s^{-1}.

Wien's law and Stefan's law can also be deduced by thermodynamical reasoning and from them it can be shown that if $E_\lambda \, d\lambda$ is the energy per unit volume of the radiation for all wavelengths between λ and $\lambda + d\lambda$, then

$$E_\lambda = \frac{1}{\lambda^5} f(\lambda T)$$

where f is a function to be determined.

In an attempt to explain the observed spectral distribution Rayleigh had postulated that the atoms in a solid oscillate due to their thermal energy and emit electromagnetic waves of the same frequency as the oscillations. By classical mechanical arguments he deduced that

$$E_\lambda = 8\pi T \lambda^{-4}$$

This is found to be approximately true for long wavelengths but is wrong for short wavelengths since the equation predicts that E_λ tends to an infinite value as the wavelength approaches zero.

The failure of classical mechanics to give the correct solution to the problem led Planck to the concept of the quantum in 1900. He postulated that an oscillator could emit or absorb energy in the form of electromagnetic radiation only in integral multiples of $h\nu$ where ν is the frequency of the oscillator and h is a constant.*

The smallest possible energy change is then

$$E = h\nu$$

and the energy would not be emitted continuously, but in bundles, each of magnitude $h\nu$. These bundles or packets are called *quanta* or *photons*. h is known as *Planck's constant* and has the value $6 \cdot 625 \times 10^{-34}$ J s.

With this restriction on energy emission, the energy distribution would be given by

$$E_\lambda = \frac{8\pi hc \lambda^{-5}}{e^{hc/k\lambda T} - 1}$$

This expression, *Planck's radiation formula*, also agrees with the experimentally determined curves shown in Figure 2.1.

Further evidence for the existence of photons is given by the phenomenon of photoelasticity which is discussed in the next section.

2.3 Photoelectric effect

When light of sufficiently short wavelength (i.e. sufficiently high photon energy) falls on to certain materials, e.g. ultraviolet light on to zinc, electrons are emitted. This is the principle utilized in some types of photocell.

* See Section 3.6.

It is found that for incident radiation of constant frequency, the rate of electron emission is proportional to the light intensity, as might be reasonably expected. But for a constant intensity of incident light, the emission varies as the frequency of the incident light is changed, there being a limiting frequency below which no electrons are produced. This frequency, which has a specific value for each material, is called the *threshold frequency*. Obviously, a certain quantity of energy has to be given to an electron to remove it from the material, but according to the wave theory of light, it should be possible to transmit that quantity of energy at any frequency, not merely at those above the threshold frequency.

In 1905, Einstein proposed an explanation which was that the energy bundles or quanta proposed by Planck continued as bundles throughout their lives and each carried the energy $h\nu$. Only those quanta or photons with sufficient energy to remove an electron could cause the photoelectric effect. Even with very low intensities of incident radiation, electrons are still emitted, showing that the energy can be concentrated locally. The minimum photon energy needed is known as the *photoelectric work function* for that material. Any excess energy that the photon possesses can be given to the electrons as kinetic energy.

This has been confirmed by experiments which showed that for incident light of some higher frequency than the minimum, the emitted electrons have some kinetic energy, the maximum value of which equals the difference between the energy of the incident photon and the work function, i.e.

$$\tfrac{1}{2} m v_{\max}^2 = h\nu - h\nu_0$$

ν_0 being the threshold frequency.

This further supported the viewpoint that light and other forms of electromagnetic radiant energy consist of pulses of electromagnetic waves, called photons, which are emitted or absorbed as discrete units, each having an energy $h\nu$, where ν is the frequency of the electromagnetic waves.

The spectrum of electromagnetic radiation is given in Figure 2.2 and shows the names by which different portions are commonly known. It also shows the relationships of frequency (and wavelength) to photon energy, the energies being quoted in *electron-volts* (eV). This unit is appropriate in size for processes of an atomic scale. The electron-volt is the energy change of a particle with a charge equal to that of an electron when it traverses a potential difference of 1 volt ($1 \text{ eV} = 1.602 \times 10^{-19}$ J).

2.4 Bohr–Rutherford atom

The concept of an atom as comprising a central positively-charged nucleus surrounded by negative charges was proposed by Rutherford in 1911, following the analysis of experiments on the scattering of α particles (which were known to have large mass relative to electrons and to be positively charged) in passing

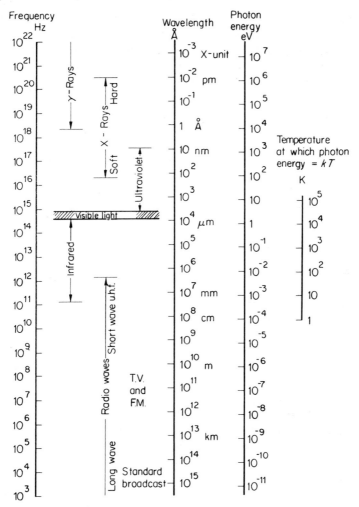

Figure 2.2. Spectrum of electromagnetic radiations in terms of frequency, wavelength and photon energy

through material. While α particles are completely absorbed by small thicknesses of metals, very thin foils let some through and also deflect some from the original direction, even to quite large angles. The observed deflections can be explained only by the existence of a repulsive electrostatic force of such magnitude that the α particle would have to approach within 10^{-14} m the positive charge in the atom. Hence the positive charge must be concentrated, together with most of the mass, in a region of about 10^{-14} m diameter, i.e. very small compared with

10^{-10} m, the diameter of the atom. The negative charges surrounding the nucleus were identified as electrons.

Study of the lines in the spectrum of light emitted by a hydrogen discharge tube had shown that their frequencies formed a definite series, the wavelengths being given by

$$\frac{1}{\lambda} = R_H \left(\frac{1}{m^2} - \frac{1}{n^2} \right) \tag{2.1}$$

where m can take any positive integral value and n can take integral values from $m + 1$ upward. R_H is known as the *Rydberg constant* and has the value 1.09736×10^7 m^{-1}.

Bohr found that the Rutherford model of an atom could explain the observed spectra of hydrogen only if certain assumptions of a quantum nature were made. These were:

a. The electrons exist only in stable circular orbits of fixed energy, the angular momentum of an electron in an orbit being an integral multiple of $h/2\pi$.
b. An electron will emit or absorb energy only when making a transition from one to another possible orbit.

Any energy given out would take the form of a photon, the energy $h\nu$ of which equalled the difference in the energies of the two orbits.

Nuclei are now known to be composed of *protons*, each with a charge equal to that of an electron, but of opposite sign, and *neutrons* which are uncharged. The number of neutrons is usually equal to or greater than the number of protons. As both proton and neutron have masses approximately 2000 times that of an electron (see Appendix 2) they may be considered to a first approximation to be at rest when dealing with the motion of the electrons around them. Actually, both electrons and nucleus must move, both orbiting about the centre of gravity.

Using the Bohr postulates and the classical laws of mechanics and of electrostatic attraction between unlike charges, the possible energy values for an electron moving around a nucleus can be calculated. The changes of energy when an electron makes a transition from one possible state to another are equal to those of photons with wavelengths given by equation (2.1).

In a hydrogen discharge tube, collisions by moving electrons with hydrogen atoms transfer electrons of the atoms from their lowest energy stable states to those of higher energy. An electron so displaced can then return to the innermost orbit by a single jump or a series of jumps. The possible energy levels of a hydrogen atom are shown in Figure 2.3 together with some possible jumps and the wavelengths of the photons emitted.

A derivation of the possible energy states in terms of wave mechanics will be given in Section 3.5.

Wave-particle Duality

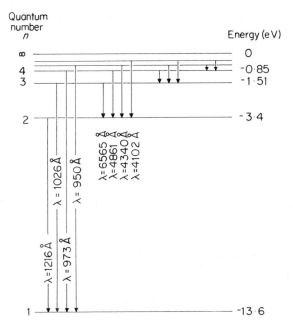

Figure 2.3. Electron energy levels in the hydrogen atom and some possible wavelengths in emission spectra

2.5 Compton effect

Further evidence of the photon nature of electromagnetic waves is shown in the scattering of X-rays by particles.

Einstein, by his special theory of relativity, showed that the mass m of a body which is moving at a velocity v would appear to an observer at rest to be greater than its rest mass m_0 and is

$$m = m_0/\sqrt{1 - (v^2/c^2)} \qquad (2.2)$$

where c is the velocity of light. This result has been confirmed by a wealth of experimental evidence. The difference between m and m_0 becomes insignificant when v is much smaller than c, in which case the relationships become those of Newtonian mechanics.

Also the kinetic energy is

$$\text{K.E.} = mc^2 - m_0c^2$$

which can be shown to equal $\tfrac{1}{2}m_0v^2$ when $v \ll c$. The quantity m_0c^2 is equivalent to an energy and is known as the *rest-mass energy*. The quantity mc^2 is called the total energy. In nuclear reactions, neither mass nor energy (as understood in the Newtonian mechanics sense) is conserved separately, but the two can interchange according to the relationship

$$E = mc^2 \tag{2.3}$$

Since mass and energy are so related, it is reasonable to consider that photons, which possess energy, should also have mass.

The energy of a photon is

$$E = h\nu = hc/\lambda$$

and its velocity is c. It will have no rest mass, but can be considered as having a moving mass given by

$$m_p = E/c^2 = h\nu/c^2$$

Also it behaves as if it has momentum p_p given by

$$p_p = h\nu/c = h/\lambda$$

Compton successfully treated the scattering of a photon by an electron as a simple collision problem with conservation of mass-energy and conservation of momentum as follows.

Assume that the photon collides with an electron initially at rest. In general, the impact is not head-on, so that the photon is deflected through an angle α and the electron moves off at an angle β to the original photon path and on the side away from the photon as shown in Figure 2.4.

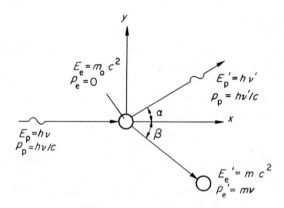

Figure 2.4. Compton scattering

If the electron has a velocity v after the impact, then its total E'_e and momentum p'_e are

$$E'_e = mc^2$$

$$p'_e = mv$$

Wave-particle Duality

where m has the value given by equation (2.2). The energy of the photon also changes, so that if v' is its new frequency, the energy and momentum are hv' and hv'/c, respectively.

Applying the law of conservation of mass-energy

$$hv + m_0c^2 = hv' + mc^2 \qquad (2.4)$$

Applying the law of conservation of momentum in directions parallel and perpendicular to the original path gives

$$hv/c + 0 = hv'/c \cos \alpha + mv \cos \beta \qquad (2.5)$$

and

$$0 = hv'/c \sin \alpha - mv \sin \beta \qquad (2.6)$$

By rearranging equations (2.5) and (2.6) to put the terms containing β on one side and by squaring and adding,

$$m^2v^2 = \frac{h^2}{c^2}(v^2 - 2vv' \cos \alpha + v'^2)$$

Substituting for v from equation (2.2) gives

$$c^2(m^2 - m_0^2) = \frac{h^2}{c^2}(v^2 - 2vv' \cos \alpha + v'^2) \qquad (2.7)$$

By rearranging equation (2.4) and squaring,

$$m^2 = m_0^2 + \frac{2hm_0}{c^2}(v - v') + \frac{h^2}{c^4}(v - v')^2 \qquad (2.8)$$

Eliminating m from equations (2.7) and (2.8)

$$c\left(\frac{1}{v'} - \frac{1}{v}\right) = \frac{h}{m_0c}(1 - \cos \alpha)$$

or, substituting the wavelengths of the photon before and after impact,

$$\lambda' - \lambda = \frac{h}{m_0c}(1 - \cos \alpha)$$

This relationship, which has been confirmed by experiment, shows that the wavelength of the scattered photon is greater than that of the incident photon by a quantity which is a function of the angle of scattering α.

The Compton scattering is significant only when the incident photons have energies of 0·1 MeV and above (i.e. only for hard X-rays and γ-rays).

2.6 de Broglie's hypothesis

Following the establishment of the dual nature of light, de Broglie, in 1924, postulated that this wave–particle duality might be characteristic of matter in general. Starting from the results of relativity theory, he showed that the wavelength that would be associated with a particle of matter would be given by

$$\lambda = h/mv$$

mv being the momentum of the particle. Thus for both photon and particle the momentum is h/λ.

2.7 Electron diffraction

Einstein pointed out that if de Broglie's hypothesis were true, then an electron beam should be capable of being diffracted by crystals just as X-rays are.

The first verification of this was found by Davisson and Germer who, in 1927, were investigating the scattering of electrons by nickel. When the piece of nickel used happened to be a single crystal, they found effects that could only be attributed to the diffraction of waves. Later, G. P. Thomson also obtained diffraction patterns by sending beams of electrons through thin sheets of various metals. From the crystal structures of the metals—which were already known from X-ray diffraction work—the wavelengths of the electron waves were determined. The energy of the electrons in a beam is given by the product of the electronic charge and the voltage V used to accelerate the electrons

$$\tfrac{1}{2}mv^2 = Ve$$

Hence the momentum and the associated de Broglie wavelength can be calculated

$$\lambda = \frac{h}{mv} = \frac{h}{\sqrt{2mVe}}$$

Substitution of values of the constants h, m and e in this equation gives the wavelength in ångströms as very nearly $(150/V)^{1/2}$, where V is in volts. For example the wavelength of a 15 kV electron is 0·1 Å. The values calculated in this way were found to agree closely with the measured values.

While these considerations were first made for electrons, they apply equally to protons, neutrons and any massive particles.

2.8 Representation of a photon or particle by a wave

The preceding discussion has shown that both electromagnetic radiation and matter, e.g. electrons, exhibit this dual nature—corpuscular and wave-like.

Wave-particle Duality

We must now consider how a pulse or packet of waves can represent a particle.

Several cycles are necessary to define the wavelength, but this immediately implies a certain lack of definition of the position of the particle.

Consider a train of waves in which the *wave function* (the term that will be used for the variable, whether it is the magnitude of the electric or magnetic vector in the case of electromagnetic waves, or some, as yet, unknown quantity in particle waves) varies in a sinusoidal manner with time. The wave function can be represented by

$$\psi = \psi_0 \sin 2\pi(x/\lambda - vt)$$

where t is the time and x is distance measured in the direction of propagation of the wave.

It will be convenient to introduce the symbol $k \, (= 2\pi/\lambda)$ which is known as the *wave number*, so that the wave function becomes

$$\psi = \psi_0 \sin(kx - 2\pi vt)$$

Such a wave (Figure 2.5) is of constant amplitude and infinitely continuous in both time and distance. Its frequency is defined exactly, but it fills all space and cannot be said to have a time of arrival at any point.

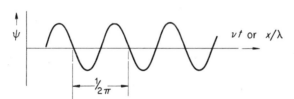

Figure 2.5. Travelling wave showing ψ as a function of time t and distance x

A photon or particle is not of infinite extent and so the wave pulse that represents it must have a beginning and an end, i.e. it is a train of waves of finite length. It is therefore an *amplitude-modulated wave*, the frequency or wavelength being constant along it but the amplitude decreasing to zero at each end. The methods of Fourier analysis show that a modulated wave is the sum of a series of constant amplitude waves. As a simpler example consider two waves of equal amplitudes but with slightly different frequencies (or wave numbers). The waves passing any point can be represented by $\tfrac{1}{2}A \sin 2\pi t(v_0 + v_1)$ and $\tfrac{1}{2}A \sin 2\pi t(v_0 - v_1)$ where $v_0 \gg v_1$.

The sum of the two equals

$$A \sin 2\pi v_0 t \cos 2\pi v_1 t$$

which is a wave of frequency v_0 modulated at a lower frequency v_1 and hence

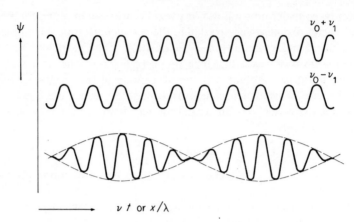

Figure 2.6. Addition of waves with frequencies $(v_0 + v_1)$ and $(v_0 - v_1)$ to give carrier wave of frequency v_0 modulated at frequency v_1

represents a train of waves of particular shape (Figure 2.6), but it is still continuous, and so unsuitable for our purpose.

We shall have to analyse a single wave packet into a more complicated series of waves which, beyond the spread of the wave packet, interfere completely and sum to zero. It is possible to do this for any form of amplitude modulation we may choose.

Consider, as an example, the wave group which at an instant of time is given by

$$\psi = e^{-x^2/\sigma^2} \sin k_0 x$$

This is a wave of constant wave number k_0 the amplitude of which is modulated by the function $\exp(-x^2/\sigma^2)$, a simple function which has the value 1 when $x = 0$ and tends rapidly to zero as x increases to values greater than σ. The form of the group, which is shown in Figure 2.7, approximates to a wave packet of finite length.

As will be seen later, the effect of a wave pulse is measured in terms of its energy, which is proportional to ψ^2. Within the range $-\sigma < x < +\sigma$ the integrated value of ψ^2 for this form of pulse is 0·84 of the value integrated over all values of x. That is one can say that the effective width of the pulse is approximately given by $\Delta x = 2\sigma$.

Fourier analysis of this will give a series of continuous waves, i.e. each of infinite extent and constant amplitude which sum to give the wave group above. The individual amplitudes are given in the equation

$$\psi = \int_{-\infty}^{\infty} \tfrac{1}{2}\pi^{-1/2} \sigma \, e^{-(k-k_0)^2 \sigma^2/4} \sin kx \, dk$$

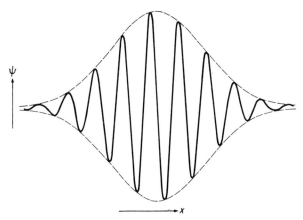

Figure 2.7. Wave packet $\psi = e^{-x^2/\sigma^2} \sin k_0 x$

i.e. there is an infinite series of waves grouped around the wave number k_0, the amplitudes of which decrease, at first slowly, and then more rapidly as the wave number moves away from k_0. The amplitude is again small except in the range bounded by $(k - k_0)\sigma/2 = \pm 1$. Hence, the spread of the wave number is approximately $\Delta k = 4/\sigma$.

Hence

$$\Delta x \cdot \Delta k \approx 8$$

Now it will not be possible to define the position of the photon or particle to a greater precision than Δx. Likewise, it will not be possible to define the wave number more precisely than Δk.

For a particle, $k = 2\pi/\lambda = 2\pi m v/h = 2\pi p/h$, where p is the momentum. Hence for a possible error Δk in k there is a possible error Δp in the momentum p given by

$$\Delta p = h\Delta k/2\pi$$

so that

$$\Delta x \cdot \Delta p \approx 8h/2\pi \approx h$$

2.9 Heisenberg's uncertainty principle

The above argument shows that a wave packet which represents a particle can simultaneously define the position and momentum to accuracies which are given by

$$\Delta x \cdot \Delta p \approx h$$

This is a statement of *Heisenberg's uncertainty principle*, that it is impossible to determine accurately the position and momentum of a particle simultaneously. The experimental refinements which are necessary to measure one of these quantities more precisely are of the type that make any measurement of the other quantity less accurate.

The hypothetical 'γ-ray microscope' which was first discussed by Heisenberg will give a better understanding of this. Suppose a beam of electrons is travelling in the x direction and we wish to find the position of an electron at some instant. We imagine that this can be done with a microscope, the beam of electrons being illuminated so that some photons scattered by the electrons enter the microscope (Figure 2.8). The precision with which the position can be observed

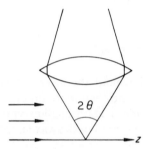

Figure 2.8. Hypothetical γ-ray microscope

is limited by the wavelength of the light and the aperture of the lens. If θ is the half angle of the cone of light and λ is the wavelength, the precision of measurement, or resolving power of the microscope is approximately $\Delta x = \lambda/2 \sin \theta$. Because $\sin \theta$ cannot exceed one and in practice must be somewhat less than one, the resolving power is limited by the wavelength λ. The greatest accuracy will therefore be obtained by using waves with the smallest value of λ, i.e. γ-rays.

A photon entering the microscope after being scattered by an electron will have changed direction and any such change of direction will have involved an exchange of momentum between the photon and the electron. The direction of the photon as it enters the microscope is not known, its momentum in the x direction may have been any value between $-(h/\lambda) \sin \theta$ and $+(h/\lambda) \sin \theta$. Suppose that the incident beam of photons is along the z direction, i.e. perpendicular to the plane of the diagram in Figure 2.9. The x component of momentum given to the electron by a photon which is scattered into the microscope may be anything between the limits given and is equal and opposite to the x component of momentum given to the electron, so that the uncertainty in the final momentum of the electron is

$$\Delta p_x = \frac{2h}{\lambda} \sin \theta$$

Wave-particle Duality

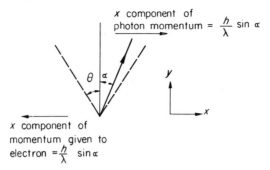

Figure 2.9. Photon scattered by electron. There is a momentum transfer of $(h/\lambda) \sin \alpha$ in the x direction

That is the more accurately we determine the electron position by decreasing λ, the larger is the uncertainty in the momentum of the electron.

$$\Delta p_x \cdot \Delta x = \left(\frac{2h}{\lambda} \sin \theta\right)\left(\frac{\lambda}{2 \sin \theta}\right) = h$$

In Newtonian mechanics, one is used to considering exact definitions of position and velocity (or momentum) at any instant, or the exact value of the energy of a body or system at a particular time. This is satisfactory unless one is dealing with particles as small as those on the atomic scale.

If, for example, the position is defined to an accuracy of 10^{-6} m, then the momentum is defined to an accuracy of $6 \cdot 6 \times 10^{-28}$ N s. For a golf ball which has a mass of 0·05 kg, this means a velocity accuracy of about 13×10^{-26} m s^{-1} which is obviously well below the practical limits of measurements. However, for an electron of mass 10^{-30} kg, the velocity would have an uncertainty of 700 m s^{-1}. Hence it is only when one is considering particles of atomic or sub-atomic size that the uncertainty principle is significant.

Because sharply defined values of position and momentum cannot be found simultaneously, it would not appear possible to develop a systematic mechanics of electrons and other atomic particles. However, by dealing only with the probabilities of position and momentum lying within certain ranges, a complete and exact mechanics can be worked out which will predict every property which can be observed experimentally.

A similar argument carried out on a basis of time and energy gives uncertainties in these quantities which are related by

$$\Delta t \cdot \Delta E \approx h$$

The longer that an electron exists in a certain energy state, i.e. the greater Δt, the more exactly is the energy of that state defined, such as an electron in a stable orbit in an atom. For a short-lived metastable state, Δt is small so that

ΔE is larger and the energy value of the state is less closely defined and any spectral lines associated with the state will have less precise values for their wavelengths.

2.10 Simple wave equations

Having discussed the representation of an electron by a wave group, we shall proceed to a study of electron behaviour by considering the behaviour of waves in various situations. First let us consider a simple case of wave motion which relates to the vibrations of a stretched string.

Let m be the linear density of the string, i.e. its mass per unit length, which is not necessarily constant but may vary with x, the distance along the string. Let there be a longitudinal tension T in the string.

Consider a small element AB (Figure 2.10) of the string of length δx and let y be the lateral displacement of AB from the equilibrium position. The tension

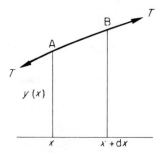

Figure 2.10. Force of an element of vibrating string

at A has a transverse component which exerts a force $T(dy/dx)$ on the element towards the undisplaced position and at B the transverse force is

$$T\left[\frac{dy}{dx} + \frac{d^2y}{dx^2}\delta x\right]$$

in the opposite direction. There will be a net force of $T(d^2y/dx^2)\delta x$ on the element which will cause a lateral acceleration d^2y/dt^2 given by

$$T\frac{d^2y}{dx^2}\delta x = m\,\delta x\frac{d^2y}{dt^2}$$

or

$$T\frac{d^2y}{dx^2} = m\frac{d^2y}{dt^2} \qquad (2.9)$$

This is the wave equation for the vibrating string.

Wave-particle Duality

Now suppose a wave of frequency v and wavelength λ is passing along the string, i.e.

$$y = y_0 \sin 2\pi(x/\lambda - vt) \tag{2.10}$$

Then each point is moving with a simple harmonic motion of frequency v and hence the motion of any point satisfies the differential equation

$$\frac{d^2y}{dt^2} + 4\pi^2 v^2 y = 0 \tag{2.11}$$

From equations (2.9) and (2.11),

$$\frac{d^2y}{dx^2} + 4\pi^2 v^2 \frac{m}{T} y = 0 \tag{2.12}$$

which is satisfied by equation (2.10) if

$$\lambda = \frac{1}{v}\sqrt{\frac{T}{m}}$$

Hence the wave velocity is given by

$$v(= \lambda v) = \sqrt{T/m}$$

Equation (2.12) can also be written in the form

$$\frac{d^2y}{dx^2} + k^2 y = 0$$

where $k (= 2\pi/\lambda)$ is the wave number at any point for waves of a given frequency. If the density of the string is not uniform, then k is not constant, but a function of x. Suppose that at a certain point the density changes abruptly from m_1 to m_2. The displacement y must be continuous at that point. From equation (2.12)

$$\frac{dy}{dx} = -\frac{4\pi^2 v^2}{T} \int m y \, dx$$

and hence, even if m changes abruptly, dy/dx must be continuous, i.e. there will be no kink in the curve.

Suppose that we now have a finite length l of the string, the ends being fixed. The displacement y of a point will be a function of x and of time. A wave travelling along the string will be reflected each time it meets a fixed end. If the frequency of the wave is correct, then the waves will reinforce and give a *standing wave* in which each point vibrates with the same frequency and with an amplitude that remains constant. The possible frequencies that satisfy this condition are obtained by stipulating the end conditions. If $y = 0$ when $x = 0$ and when $x = l$, then the only possible solutions are sine functions with nodes at each end of the string, i.e.

$$n\lambda/2 = l$$

where n can take all positive integral values. Hence

$$v = \frac{1}{\lambda}\sqrt{\frac{T}{m}} = \frac{n}{2l}\sqrt{\frac{T}{m}}$$

or, the frequency can only have certain values. The forms of the vibration for $n = 1, 2$ and 3 are shown in Figure 2.11.

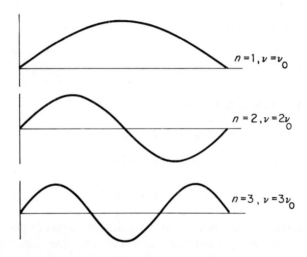

Figure 2.11. First three vibration modes of taut string

2.11 Wave representation of a beam of electrons

If a wave is to represent a beam of electrons, it must travel in the direction of the beam and its amplitude must be related to the density of electrons in the beam.

In fact, because of the uncertainty principle, we cannot strictly talk of the density of electrons, but rather the *probability of finding an electron* at any given point. The wave intensity $|\psi|^2$ is taken to be proportional to this probability and it is convenient to choose the units of the wave function ψ so that the intensity is equal to the probability.

The wave function may be a complex quantity, i.e.

$$\psi = f + ig$$

Wave-particle Duality

where f and g are real functions of position and time. Then the intensity of the wave is the product of ψ and its complex conjugate $\psi^* (= f - ig)$, i.e.

$$|\psi|^2 = \psi\psi^*$$
$$= (f + ig)(f - ig)$$
$$= f^2 + g^2$$

At a position where $|\psi|^2$ varies with time, the probability of the presence of an electron varies. However, the average value of $|\psi|^2$ with respect to time is the probability of finding an electron at that position at any randomly selected time, or the average charge density at the point. Many problems can conveniently be discussed in terms of this last interpretation.

A plane wave travelling along the x axis would have the form

$$A \sin(kx - 2\pi vt + \theta)$$

where θ is a phase angle. Hence both f and g will be functions of this form but possibly with different values of θ.

Now presumably in a beam of electrons of uniform density, the probability of finding an electron at any point along the beam should be independent of both t and x, i.e. $|\psi|^2$ should not vary with t or x. This is possible if f and g are of equal amplitude and 90° out of phase, i.e.

$$f = A \cos(kx - 2\pi vt)$$
$$g = A \sin(kx - 2\pi vt)$$

because then

$$f^2 + g^2 = A^2$$

and the wave function is

$$\psi = A \cos(kx - 2\pi vt) + iA \sin(kx - 2\pi vt)$$
$$= A e^{i(kx - 2\pi vt)}$$

A wave travelling in the opposite direction will be represented by

$$\psi = B e^{i(-kx - 2\pi vt)}$$

In each of these expressions, the time variant and the space variant are separable, so that the wave functions can be written as products in the form

$$\psi = A e^{ikx} e^{-2\pi ivt}$$

and

$$\psi = B e^{-ikx} e^{-2\pi ivt}$$

The time factor is always of the form $e^{-2\pi ivt}$.

Now these equations are solutions of a differential equation, which, following the procedure in Section 2.10, can be written in various forms:

$$\frac{d^2\psi}{dx^2} + k^2\psi = 0 \tag{2.13}$$

$$\frac{d^2\psi}{dt^2} + 4\pi^2 v^2 \psi = 0$$

or

$$\frac{d^2\psi}{dx^2} = \frac{k^2}{4\pi^2 v^2}\frac{d^2\psi}{dt^2}$$

2.12 Formulation of Schrödinger's equation

We now wish to derive the wave equation for a single particle. Let it have a mass m and be moving along the x axis while subject to a force in that direction such that its potential energy is $V(x)$. The total energy E is equal to the sum of the kinetic and potential energies, so that

$$E = \tfrac{1}{2}mv^2 + V$$

$$= \frac{p^2}{2m} + V$$

Thus

$$p = \sqrt{2m(E-V)}$$

Now by de Broglie's hypothesis, the corresponding waves would have a wavelength

$$\lambda = \frac{h}{p} = \frac{h}{\sqrt{2m(E-V)}}$$

or wave number

$$k = \frac{2\pi}{\lambda} = \left\{\frac{8\pi^2 m(E-V)}{h^2}\right\}^{1/2} \tag{2.14}$$

so that equation (2.13) becomes

$$\frac{d^2\psi}{dx^2} + \frac{8\pi^2 m(E-V)}{h^2}\psi = 0$$

If the motion of the particle is not confined to one dimension, and it is travelling in a direction which has direction cosines l, m and n, relative to coordinate axes x, y and z, then the wave function will be

$$\psi = A\, e^{i[k(lx+my+nz)-2\pi\nu t]} \tag{2.15}$$

Now

$$l^2 + m^2 + n^2 = 1$$

and so equation (2.15) is a solution of

$$\nabla^2\psi + k^2\psi = 0 \tag{2.16}$$

where the Laplacian operator ∇^2 has the meaning

$$\frac{\partial^2}{\partial x^2} + \frac{\partial^2}{\partial y^2} + \frac{\partial^2}{\partial z^2}$$

Equation (2.16) can be satisfied by any number of wave functions of the type given in equation (2.15), all of which have different directions and amplitudes, but all having the same wave number k, which is given by

$$k^2 = \frac{8\pi^2 m(E-V)}{h^2}$$

The differential equation (2.16) then becomes

$$\nabla^2\psi + \frac{8\pi^2 m}{h^2}(E-V)\psi = 0$$

This is Schrödinger's equation in the form in which it is independent of time, i.e. analogous to equation (2.12) in Section 2.10.

Again as in Section 2.10, we assume that this equation applies even when V varies with position, i.e. the value of k will change with position.

2.13 Conditions for solution of Schrödinger's equation

To solve Schrödinger's equation in any particular case, the value of V must first be expressed as a function of position.

Then there are three important conditions which restrict the possible solutions:

a. The integral $\iiint |\psi|^2\, dx\, dy\, dz$, where the integration is taken over all space and represents the total probability of finding the electron anywhere, must be finite.
b. ψ must be continuous and single-valued.
c. $\partial\psi/\partial x$, $\partial\psi/\partial y$ and $\partial\psi/\partial z$ must be continuous.

In some problems solutions are obtainable only for certain values of E which are called *eigenvalues*. The corresponding expressions for ψ are called *eigenfunctions*. Some particular cases are considered in Sections 3.3–3.5.

2.14 Normalization

$|\psi|^2$ has been defined as being proportional to the probability of finding an electron in a given space interval. It would be convenient if the arbitrary constant in the solution for ψ were adjusted so that $|\psi|^2 \, \delta x \, \delta y \, \delta z$ is the actual probability of finding the electron in the volume element $\delta x \, \delta y \, \delta z$. Then for a single electron problem, the probability of finding the electron somewhere in all space must be unity, i.e.

$$\iiint |\psi|^2 \, dx \, dy \, dz = 1$$

When this condition is satisfied by a suitable value of A in equation (2.15) the wave function is said to be *normalized*.

Questions for Chapter 2

1. Calculate the wavelengths of the spectral lines of atomic hydrogen corresponding to electrons jumping from the $n = 2$ to $n = 1$ and $n = 4$ to $n = 3$ levels.

2. The energy levels in singly ionized helium atoms have energies four times those of the corresponding levels in hydrogen atoms. Calculate the second ionization potential in helium, i.e. the energy, in eV, which is needed to remove the remaining electron from the He$^+$ ion.

3. A diffraction experiment shows that a certain beam of electrons exhibits a wavelength of 3·6 Å. Calculate the velocity of the electrons and their energy in electron-volts.

4. Calculate the wavelength of a billiard ball of mass 0·2 kg moving with a velocity of 2·5 m s^{-1}.

5. Calculate for an electron which has been accelerated by a potential difference of 6500 V (*a*) its momentum, (*b*) its de Broglie wavelength, (*c*) its wave number.

6. If an excited state of an atom has a lifetime of 2×10^{-4} s, to what accuracy (in electron-volts) can the energy of this excited state be determined?

7. Light, of wavelength 5889 Å, is incident on a photocathode having a work function of 1·2 eV. Calculate the maximum velocity of the emitted electrons.
[MST]

8. Given that electrons, each with momentum p, have properties which correspond to those of waves with wavelength $\lambda = h/p$, explain the reasoning which leads to the one-dimensional, time-independent form of Schrödinger's

equation

$$\frac{d^2\psi}{dx^2} + \frac{8\pi^2 m}{h^2}(E - V)\psi = 0$$

[MST]

CHAPTER THREE

Wave Mechanics Applied to Single Electrons

3.1 Motion of a beam of electrons through potential jumps

It was shown in Section 2.11 that a beam of electrons can be represented by a wave equation of constant amplitude

$$\psi = A\, e^{i(kx - 2\pi vt)}$$

The value of k is related to the kinetic energy of the electrons, so that if the beam moves through a region of variable potential, k will vary along the path. Two particular cases which are of importance in the operation of electronic devices will be considered here.

3.1.1 Effect of a potential jump

Consider a beam of electrons travelling in the x direction, each electron having a total energy E. Suppose that the path of the beam lies in two different regions, the potential energy of an electron having a different constant value in each, being V_0 for $x < 0$ and V_1 for $x > 0$ (Figure 3.1) which is referred to as an energy step or a potential jump at $x = 0$.

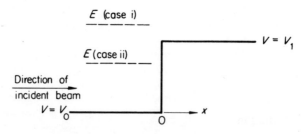

Figure 3.1. Energy step of height $(V_1 - V_0)$

Now if the electrons are travelling towards the jump in the region where $x < 0$, the potential energy V_0 must be less than the total energy E, the difference being the kinetic energy, so that the velocity is

$$v_0 = \sqrt{2(E - V_0)/m} \qquad (3.1)$$

If $E > V_1$, the electrons will have kinetic energy in the second region and continue to travel with a velocity

$$v_1 = \sqrt{2(E - V_1)/m} \qquad (3.2)$$

On the other hand, if $E < V_1$, the electrons would have negative kinetic energy in the region, which is impossible. Hence on classical considerations, the electrons would not enter this region, but be reflected at $x = 0$.

When these two cases are studied in terms of wave mechanics, we get results which differ from those just given.

Case (i) $E > V_1$. For $x < 0$, Schrödinger's equation can be written as

$$\frac{d^2\psi}{dx^2} + \alpha^2\psi = 0$$

where

$$\alpha^2 = 8\pi^2 m(E - V_0)/h^2 \qquad (3.3)$$

and for $x > 0$

$$\frac{d^2\psi}{dx^2} + \beta^2\psi = 0$$

where

$$\beta^2 = 8\pi^2 m(E - V_1)/h^2 \qquad (3.4)$$

Now suppose that the incident wave is of unit amplitude, i.e. it is given by

$$\psi = e^{i(\alpha x - 2\pi v t)}$$

There is a possibility that some of the incident beam may be reflected and some may continue into the region for $x > 0$. Let the amplitudes of the reflected and transmitted portions be A and B, respectively.

Then the wave function for $x < 0$ must include both incident and reflected waves and will be

$$\psi = (e^{i\alpha x} + A e^{-i\alpha x}) e^{-2\pi i v t} \qquad (3.5)$$

while the wave function for $x > 0$ will be

$$\psi = B e^{i(\beta x - 2\pi v t)}$$

It was stated in Section 2.13 that both ψ and $d\psi/dx$ must be continuous at all points. Applying these conditions at $x = 0$ gives

$$1 + A = B \tag{3.6}$$

and

$$\alpha(1 - A) = \beta B \tag{3.7}$$

From equations (3.6) and (3.7)

$$A = \frac{\alpha - \beta}{\alpha + \beta}$$

and

$$B = \frac{2\alpha}{\alpha + \beta}$$

While A and B cannot be expressed as simple functions of E, V_0 and V_1, they are simply related to the electron velocities in the two regions.

From equations (3.1) and (3.3)

$$v_0 = \alpha h/2\pi m$$

Similarly, from equations (3.2) and (3.4)

$$v_1 = \beta h/2\pi m$$

Hence

$$A = \frac{v_0 - v_1}{v_0 + v_1}$$

and

$$B = \frac{2v_0}{v_0 + v_1}$$

Now the density of electrons is $|\psi|^2$ which has the value unity in the incident beam, $|A|^2$ in the reflected beam and $|B|^2$ in the transmitted beam. The number of electrons N_i in the incident beam crossing unit area in unit time is the product of the electron density and velocity, i.e.

$$N_i = \alpha h/2\pi m$$

The corresponding numbers in the reflected and transmitted beams, respectively, will be

$$N_r = \frac{\alpha h}{2\pi m}|A|^2$$

$$= N_i \left(\frac{v_0 - v_1}{v_0 + v_1}\right)^2$$

Wave Mechanics Applied to Single Electrons

and

$$N_t = \frac{\beta h}{2\pi m}|B|^2$$

$$= N_i \frac{4v_0 v_1}{(v_0 + v_1)^2}$$

It can be seen that

$$N_i = N_r + N_t$$

so that, as would be expected, the total number of particles remains constant. However, as the height of the potential jump is increased, causing v_1 to decrease, an increasing proportion of the electrons are reflected. This contrasts with the classical solution according to which, none would be reflected.

If the electron beam were to consist of a single electron, then, obviously, part of it could not be reflected. In such a case, the probability aspect must be considered. There will be reflected and transmitted wave functions of intensities $|A|^2$ and $|B|^2$, respectively. But if one performs an experiment to look for the electron that has reached the jump, then it will be found in one or the other region only, the probability of finding it in the region $x < 0$ being $|A|^2$ and in the region for $x > 0$ being $|B|^2$.

It may be noted in this case that A and B are real quantities and B is always positive. That is, the components of the transmitted wave function are in-phase with the components of the incident wave function. The reflected wave function is in-phase or completely out-of-phase with the incident wave function, depending on whether v_0 is greater or less than v_1, i.e. whether V_0 is greater or less than V_1.

Case (ii) $E < V_1$. In this case, $(E - V)$ is negative for $x > 0$, so that it is convenient to write Schrödinger's equation in the form

$$\frac{d^2\psi}{dx^2} = \gamma^2 \psi$$

where

$$\gamma^2 = 8\pi^2 m(V_1 - E)/h^2 \tag{3.8}$$

The general solution of this equation is of the form

$$\psi = (C\,e^{\gamma x} + D\,e^{-\gamma x})\,e^{-2\pi i v t} \tag{3.9}$$

Now as $x \to \infty$ so also does $e^{\gamma x}$, which would mean an infinite probability of finding the electron at infinity. This is obviously impossible, so that $C = 0$. Also as $x \to \infty$, $e^{-\gamma x} \to 0$ and hence is an acceptable solution.

We take the incident and reflected wave to be given by equation (3.5) as in case (i) and the transmitted wave to be

$$\psi = D e^{(-\gamma x - 2\pi i v t)}$$

The boundary conditions that ψ and $d\psi/dx$ are continuous at $x = 0$ will give

$$1 + A = D$$

and

$$i\alpha(1 - A) = -\gamma D$$

from which

$$A = \frac{\alpha - i\gamma}{\alpha + i\gamma} \tag{3.10}$$

The intensity of the reflected wave is

$$|A|^2 = AA^* = \frac{\alpha - i\gamma}{\alpha + i\gamma} \cdot \frac{\alpha + i\gamma}{\alpha - i\gamma}$$
$$= 1$$

i.e. it is equal to the intensity of the incident wave, or, in other words, all the electrons are reflected.

However, the wave function does not vanish abruptly at the point of reflection, but for $x > 0$ decays according to $e^{-\gamma x}$, i.e. the higher the jump, the more rapid is the decay.

If the potential were to drop again to a value less than E at some higher value of x, then the wave function would have a finite value at that point giving a continuous non-decaying function to higher values of x, i.e. some of the beam would pass through the barrier. A particular case of this is considered in the next Section.

3.1.2 *Effect of rectangular barrier*

Consider a potential barrier of width a which is bounded by two potential jumps such that

$$V_x = V_0 \quad \text{for} \quad x < 0$$
$$V_x = V_1 \quad \text{for} \quad 0 < x < a$$

and

$$V_x = V_0 \quad \text{for} \quad x > a$$

and suppose that the total energy E of electrons in a beam incident upon the barrier has a value between V_0 and V_1 (Figure 3.2).

Wave Mechanics Applied to Single Electrons

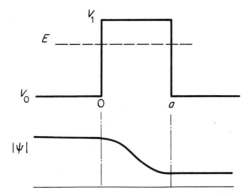

Figure 3.2. Energy barrier showing tunnel effect

The effect of the jump at $x = 0$ is that there will be an exponential decay type of wave function for $x > 0$ which will not have fallen to zero at $x = a$. At this point there will be partial reflection but for $x > a$ there will be a wave function with constant amplitude. Hence the electron beam will continue with an amplitude determined by the width of the barrier. Some of the electrons have penetrated the barrier, even though on classical arguments there should have been complete reflection. The proportion T of the incident electrons which pass the barrier is known as the *transmission coefficient* or *transparency*.

If a single electron approaches the barrier, then there is a probability equal to T that it will be found on the far side. This phenomenon of particles traversing potential barriers, which should on classical arguments reflect them, is known as the *tunnel effect*.

The case just considered is highly artificial, as in practical cases the potential variation is not an abrupt step and the problem cannot be solved by rigorous analytical methods. However, the same principle applies and the transparency of such barriers is found to be of the order of $e^{-c\gamma a}$ where γ has the meaning already defined in equation (3.8), V_1 being the height of the potential barrier, a is the width of the barrier at the level of E (see Figure 3.3) and c is a numerical constant, the value of which depends upon the shape of the barrier and is of the order of 1–2.

Figure 3.3. Variable-width energy barrier

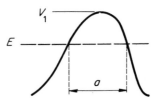

Consider the case of a particle approaching a barrier which has a height $(V_1 - E) = 1$ eV and a width $a = 1$ Å. If the particle is an electron, then

$$\gamma = [8\pi^2 m(V_1 - E)/h^2]^{1/2}$$
$$= 5 \cdot 1 \times 10^9 \text{ m}^{-1}$$

The transparency is

$$e^{-\gamma a} = e^{-0 \cdot 51}$$
$$\approx 0 \cdot 6$$

so that the electron has a high chance (60 per cent) of penetrating such a barrier.

If the particle is a proton, then

$$\gamma = 2 \cdot 2 \times 10^{11} \text{ m}^{-1}$$

and

$$e^{-\gamma a} = e^{-22}$$
$$\approx 10^{-10}$$

which is so small as to be negligible.

Electrons can pass freely through a potential barrier of a few electron-volts in height and a few angströms width, which is why an electric current will flow across the junction between two metals in contact even when the surfaces are contaminated with oxides or grease.

Only barriers of much smaller widths will have similar transparencies for massive particles (protons and neutrons). The tunnel effect is involved in the radioactive emission of α particles from the nuclei of certain isotopes and in many other nuclear reactions.

3.2 General consideration of stationary states

So far in this chapter, only beams of electrons have been considered. Of more importance are the cases where electrons are confined to a definite region by potential barriers on all sides, such as an electron in an atom or the valence electrons in a metal crystal.

Consider an electron which can move in a straight line under the action of an attractive force which is always directed to some point O on the line. As the electron moves away from O in either direction, its potential V will increase in some manner such as that shown in Figure 3.4(a). If the electron has a total energy E, then Schrödinger's equation for the wave function of the electron is

$$\frac{d^2\psi}{dx^2} + \frac{8\pi^2 m}{h^2}(E - V)\psi = 0$$

Wave Mechanics Applied to Single Electrons

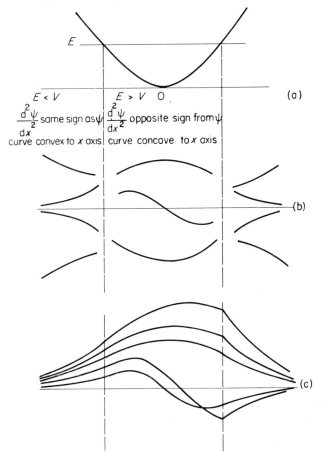

Figure 3.4. (a) Potential as function of position. (b) Possible shapes of wave function corresponding to potential well. (c) Joining of wave functions. The only permissible ones are those which join smoothly

In the region near O, for which $E > V$, then $(E - V)$ is positive so that $d^2\psi/dx^2$ and ψ have opposite signs. The curvature of the $\psi - x$ graph is approximately equal to $d^2\psi/dx^2$ so that, in this region, the graph must be concave to the x axis.

In the regions further from O, where $E < V$, then $d^2\psi/dx^2$ and ψ will have the same sign and the $\psi - x$ curves will be convex to the x axis.

The curves will therefore be restricted to certain shapes, of which possibilities are shown in Figure 3.4(b). In addition to the curvature limitations, a point of inflexion must occur when $E = V$.

Where $E < V$, the particle cannot have kinetic energy and therefore, on classical theory, could not exist. However, it was shown in Section 3.1 that there is a solution of Schrödinger's equation that can be of the form given in equation (3.9). For the same reasons as given there, the only acceptable solutions are those for which ψ tends to zero at large distances from O, i.e. the exponential decay solutions.

In the type of problem now being considered, the wave function does not represent a beam of electrons of constant strength, so that the solution of Schrödinger's equation will not necessarily be of the form e^{ikx}, which has constant amplitude, but will be of the more general form $A \cos kx + B \sin kx$ which has a varying amplitude, and where A and B may be complex.

In the central region where $E > V$, the wave number of the wave function varies as the energy E changes. If the wave function is to join smoothly with the exponential decay portions, then only certain solutions are again permissible. For example in Figure 3.5(c) various curves are shown, each of which joins smoothly with one exponential decay curve on the left, but only certain ones will also join on the right-hand side. Hence there are only certain values of the energy for which a solution is possible. This is examined further in a quantitative manner in the following sections.

3.3 Stationary states in a potential well

3.3.1 *Linear box or deep potential well*

This is an artificial problem which is a simple example of a case when Schrödinger's equation can have only certain restricted solutions. Variation of V and hence of ψ are considered in one dimension only.

Let $V = 0$ inside the box, which extends from $x = 0$ to $x = a$, and jump to ∞ at the ends as shown in Figure 3.5. If E is the energy of an electron, then for $0 < x < a$

$$\frac{d^2\psi}{dx^2} + \frac{8\pi^2 mE}{h^2}\psi = 0$$

The general solution of this will be

$$\psi = A \cos \frac{2\pi}{h}\sqrt{2mEx} + B \sin \frac{2\pi}{h}\sqrt{2mEx}$$

Outside the box, the equation becomes

$$\frac{d^2\psi}{dx^2} = \infty \psi$$

and since $d^2\psi/dx^2$ must always be finite (otherwise $d\psi/dx$ and ψ could be infinite) then $\psi = 0$, i.e. there is zero probability of an electron being in a region where $V = \infty$.

Wave Mechanics Applied to Single Electrons

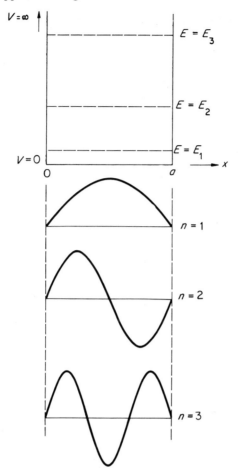

Figure 3.5. Infinitely deep one-dimensional potential box and first three wave functions

To give continuity at the boundary, the only possible solutions are those for which $\psi = 0$ both at $x = 0$ and at $x = a$. Therefore $A = 0$ and

$$\sin \frac{2\pi}{h}\sqrt{2mEa} = 0$$

The latter restriction is possible only if

$$\frac{2\pi}{h}\sqrt{2mEa} = n\pi \qquad (3.11)$$

where n is an integer.

Hence the possible solutions are

$$\psi = B \sin \frac{n\pi x}{a}$$

For $n = 1$, the wave function takes the form of a half sine wave (Figure 3.5). The probability of finding the electron is greatest at the mid-point.

For $n = 2$,

$$\psi = B \sin \frac{2\pi x}{a}$$

and the wave function is a complete cycle of a sine wave, with a node at the mid-point of the box.

From equation (3.11) we see that there is a solution only if the energy has one of certain discrete values which are given by

$$E = \frac{n^2 h^2}{8ma^2}$$

The value of n is called the quantum number of the particular wave function. For each quantum state defined by a particular quantum number, there is a specific energy level or *eigenvalue*. These eigenvalues increase in the ratios $1^2 : 2^2 : 3^2 \ldots n^2$.

For any one quantum state, the total probability of finding the electron in the linear box must be one. The value of B necessary for this normalization is found by putting

$$\int_0^a |\psi|^2 \, dx = 1$$

i.e.

$$\int_0^a B^2 \sin^2 \frac{n\pi x}{a} \, dx = \frac{aB^2}{2}$$
$$= 1$$

so that

$$B^2 = 2/a$$

and the wave function is

$$\psi = (2/a)^{1/2} \sin \frac{n\pi x}{a}$$

3.2.2 Linear box with low sides

In the previous case, because of the infinite height of the walls of the box, the probability of the electron being outside the box was zero. When the sides of the box are of limited height, then the outside region must be considered.

Wave Mechanics Applied to Single Electrons

Let the potential within the box be V_0 and that outside be V_1 and consider an electron with energy E such that $V_0 < E < V_1$ as in Figure 3.6.

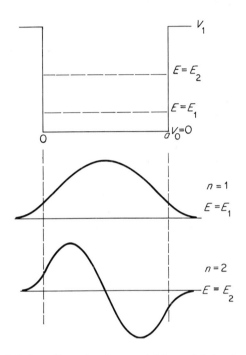

Figure 3.6. One-dimensional potential box of finite depth and first two wave functions

Defining α and γ as in Section 3.1, then within the box

$$\frac{d^2\psi}{dx^2} + \alpha^2\psi = 0$$

for which the solution is

$$\psi = A \sin \alpha x + B \cos \alpha x$$

and outside the box,

$$\frac{d^2\psi}{dx^2} = \gamma^2\psi$$

the solution of which is

$$\psi = C e^{\gamma x} + D e^{-\gamma x}$$

For $x > a$, $e^{\gamma x}$ tends to infinity as x gets larger. Hence C must be zero in this region. Similarly for negative values of x, D must be zero, so that outside the box, the wave functions must be

$$\psi = D e^{-\gamma x} \quad \text{for} \quad x > a$$

and

$$\psi = C e^{\gamma x} \quad \text{for} \quad x < 0$$

That is ψ will vary in a sinusoidal manner within the box and will decay in an exponential manner outside the box. The curves must join smoothly at $x = 0$ and at $x = a$, which places restrictions on the possible results. Again there will be only certain wave functions which are permissible, the shapes for the two lowest energies being shown in Figure 3.6.

If the electron possessed energy E which was greater than V_1, then, of course, it could escape. This is what happens in photoelectric and thermionic emission.

An alternative treatment of the linear box problem is as follows. The electron in the box will be in motion and can be represented by a wave

$$\psi = A\, e^{i(kx - 2\pi vt)}$$

where

$$k = [8\pi^2 m(E - V_0)/h^2]^{1/2}$$

When the wave reaches the edge of the box, then if $E < V_1$ so that it cannot escape, there will be complete reflection of the wave. The reflected wave will be given by

$$\psi = B\, e^{i(-kx - 2\pi vt)}$$

where the value of B will be given by equation (3.10), i.e.

$$B = A\frac{\alpha - i\gamma}{\alpha + i\gamma}$$

α and γ having the meanings stated in equations (3.3) and (3.8).

For the case of a very high-sided box, $\gamma \gg \alpha$ and so

$$B \approx -A$$

The sum of the incident and reflected waves then becomes

$$\psi = A(e^{ikx} - e^{-ikx})\, e^{-2\pi i vt}$$
$$= 2iA \sin kx\, e^{-2\pi i vt}$$

That is the combination of the incident and reflected waves form a standing wave, the amplitude of which varies in a sinusoidal manner along the box and does not vary with time.

If the amplitude of the standing wave is to be zero at each end of the box, then

$$k = \frac{\pi n}{a}$$

giving, as before,

$$E = \frac{n^2 h^2}{8ma^2}$$

as the only permitted values of energy.

3.4 Three-dimensional potential well

The previous discussion was limited to the artificial case of an electron restricted to motion in only one dimension. The general problem of the energy states of electrons in a solid, consideration of which will frequently recur in later chapters, involves motion in three dimensions, the boundaries of the solid being potential barriers.

The problem of Section 3.3 is extended to three dimensions by considering a rectangular box extending from 0 to a in the x direction, from 0 to b in the y direction and from 0 to c in the z direction.

With the potential equal to zero inside the box and infinity outside, then

$$\nabla^2 \psi + \frac{8\pi^2 m E}{h^2} \psi = 0$$

within the box.

The solutions that satisfy the boundary conditions are

$$\psi = A \sin \frac{\pi n_x x}{a} \sin \frac{\pi n_y y}{b} \sin \frac{\pi n_z z}{c}$$

where n_x, n_y and n_z are integers, each of which can have any positive value.

This is the standing wave formed by the reflection at the sides of the box of the travelling electron wave

$$\psi = A\, e^{i(k_x x + k_y y + k_z z - 2\pi v t)}$$

where

$$k_x = \frac{\pi n_x}{a}, \quad k_y = \frac{\pi n_y}{b} \quad \text{and} \quad k_z = \frac{\pi n_z}{c}$$

are the x, y and z components, respectively, of the wave vector **k**. The direction of this vector is normal to the wave front and its magnitude k equals $2\pi/\lambda$.

The energy appropriate to a solution is given by

$$E = \frac{h^2}{8m}\left\{\left(\frac{n_x}{a}\right)^2 + \left(\frac{n_y}{b}\right)^2 + \left(\frac{n_z}{c}\right)^2\right\} \tag{3.12}$$

$$= \frac{h^2}{8\pi^2 m}\{k_x^2 + k_y^2 + k_z^2\}$$

$$= \frac{h^2 k^2}{8\pi^2 m}$$

Each allowed value of energy is defined by three quantum numbers (n_x, n_y and n_z). Also each quantum number is one more than the number of nodal planes of the wave function perpendicular to the particular axis. Thus for (1, 1, 2) there is one nodal plane perpendicular to the z axis and at a distance $c/2$ from the origin.

It is possible to have two or more solutions which correspond to the same energy value. For example if the box had two sides equal ($a = b$) then the solution with quantum numbers (2, 1, 1) would have the same energy as that for (1, 2, 1). Such corresponding energy states are said to be *degenerate*.

If the box were a cube, then the three states (2, 1, 1), (1, 2, 1) and (1, 1, 2) will have the same energy and form a triple degenerate state.

At a later stage (Section 10.6) we will need to know the density of stationary states at various energy levels. We will now calculate the number of stationary states with energies below a value E.

Let the values k_x, k_y and k_z of each possible state be the coordinates of a lattice point referred to the axes x, y and z in k space. The sphere given by the equation

$$k_x^2 + k_y^2 + k_z^2 = \frac{8\pi^2 m}{h^2}E$$

will pass through any point corresponding to a quantum state with energy E and will contain within its positive octant all points which correspond to stationary states with energy less than E. The radius of the sphere is $\sqrt{8\pi^2 mE/h^2}$ and hence the volume of its octant is

$$\frac{1}{8} \cdot \frac{4\pi}{3}\left(\frac{8\pi^2 mE}{h^2}\right)^{3/2}$$

The points in k space are separated by distances π/a, π/b and π/c in the x, y and z directions, respectively, so that each point corresponds to a volume π^3/abc.

Wave Mechanics Applied to Single Electrons

Hence the number of stationary states with energy less than E is

$$\frac{1}{8} \cdot \frac{4\pi}{3} \left(\frac{8\pi^2 mE}{h^2} \right)^{3/2} \frac{abc}{\pi^3} = \frac{\pi}{6} \left(\frac{8mE}{h^2} \right)^{3/2} V$$

where V is the volume of the box. It will be seen that this value is not a function of the linear dimensions but only of the total volume.

3.5 Particle moving in a central field

This is the problem of an electron moving in a field for which the potential is a function only of the distance from a given point. A particular case of this is the electron in a hydrogen atom for which the potential energy is given by Coulomb's law

$$V(r) = -\frac{e^2}{4\pi\varepsilon_0 r} \qquad (3.13)$$

where r is the distance from the nucleus. That is it is a particular case of a three-dimensional box with sloping sides as shown in Figure 3.7, with $V \to 0$ as $r \to \infty$.

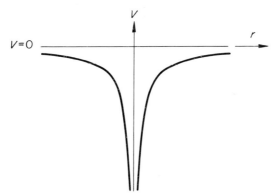

Figure 3.7. Potential well around hydrogen nucleus

Schrödinger's equation is

$$\nabla^2 \psi + \frac{8\pi^2 m}{h^2}(E - V)\psi = 0$$

If E is positive, then the electron is not confined within the box. Stationary states will exist only for $E < 0$; that is the total energy is negative relative to the arbitrary zero which has been chosen as the energy of an electron at rest at infinity.

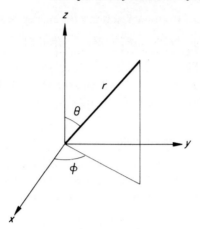

Figure 3.8. Relationship between cartesian and spherical polar coordinates

As the potential is a function of r only, i.e. spherically symmetrical, it is more convenient to express $\nabla^2 \psi$ in terms of spherical polar coordinates r, θ, ϕ. The relationship of these to rectangular cartesian coordinates is shown in Figure 3.8. Schrödinger's equation then becomes

$$\frac{1}{r^2}\frac{\partial}{\partial r}\left(r^2 \frac{\partial \psi}{\partial r}\right) + \frac{1}{r^2 \sin \theta} \frac{\partial}{\partial \theta}\left(\sin \theta \frac{\partial \psi}{\partial \theta}\right) + \frac{1}{r^2 \sin^2 \theta} \frac{\partial^2 \psi}{\partial \phi^2}$$

$$+ \frac{8\pi^2 m}{h^2}\left(E - \frac{e^2}{4\pi\varepsilon_0 r}\right)\psi = 0 \qquad (3.14)$$

The boundary conditions are

$$\left.\begin{array}{c} \psi(r, \theta, \phi) = \psi(r, \theta, \phi + 2\pi) \\ \psi(r, \theta, \phi) = \psi(r, \theta + 2\pi, \phi) \\ \lim_{r \to \infty} \psi(r, \theta, \phi) = 0 \end{array}\right\} \qquad (3.15)$$

The first two conditions are a consequence of the fact that the function must be single-valued, i.e. if either angular coordinate is changed by 2π, the value of the function must be the same. The third condition gives the exponential-type decay dealt with in Section 3.3.

A further condition to be satisfied is the normalization condition that

$$\iiint |\psi|^2 \, dV = 1 \qquad (3.16)$$

where the integral is taken to infinity in all directions.

Wave Mechanics Applied to Single Electrons

The complete solution is too lengthy to be given here but is put in some detail in Appendix 3. It is found that solutions are obtainable only for certain values of the three quantum numbers n, l and m_l.

The permissible energy levels depend only upon the principal quantum number n which can take all integral values from one upwards, the energies being

$$E = -\frac{me^4}{8\varepsilon_0^2 h^2 n^2}$$

The second quantum number l can take all integral values from 0 to $n-1$ and the third quantum number m_l can take all integral values from $+l$ to $-l$. The states corresponding to $l = 0, 1, 2, 3$ are described by the letters s, p, d, f, respectively. Thus an electron with quantum numbers $n = 3, l = 1$, is referred to as being in a $3p$ state.

For each value of l, $2l + 1$ solutions are possible, so that for each value of n, the total number of solutions is

$$\sum_{l=0}^{l=n-1} (2l + 1) = n^2$$

Thus for each energy level other than that for $n = 1$, there is degeneracy with a multiplicity of four for $n = 2$, nine for $n = 3$, etc.

The wave function for $n = 1$ (for which also $l = 0$, $m_l = 0$) is

$$\psi = \frac{1}{\pi a^3} e^{-r/a}$$

where

$$a = \frac{\varepsilon_0 h^2}{\pi m e^2}$$

The variation of ψ with r is shown in Figure 3.9. From this it would seem that

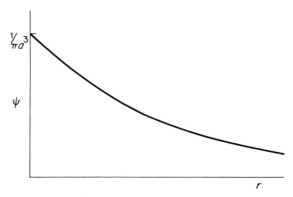

Figure 3.9. 1s wave function for hydrogen atom

the most probable position of the electron is close to the nucleus. But the product of $|\psi|^2$ at r and the volume of a thin spherical shell of radius r and thickness dr is the probability of finding the electron between a distance r and a distance $r + dr$ from the nucleus. The volume of successive shells increases more rapidly than $|\psi|^2$ decreases when r is small, so that, as may be seen from Figure 3.10, the most probable distance of the electron from the nucleus is not zero. Simple calculation shows that this most probable distance is a.

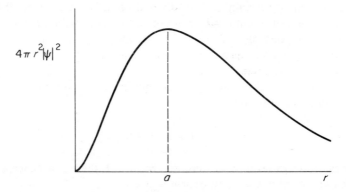

Figure 3.10. Variation of electron density with radius for 1s state in hydrogen atom

For $n = 2$, the wave functions are more complicated. When $l = 0$ and $m_l = 0$,

$$\psi = \frac{1}{2\sqrt{2\pi a^3}} e^{-r/2a}\left(1 - \frac{r}{2a}\right)$$

and when $l = 1$,

$$\psi = \frac{1}{2\sqrt{2\pi a^3}} e^{-r/2a} \frac{r}{2a} P$$

where

$$P = \cos\theta \quad \text{for} \quad m_l = 0$$

and

$$\left.\begin{array}{l} P = \sin\theta \cos\phi \\ P = \sin\theta \sin\phi \end{array}\right\} \quad \text{for} \quad m_l = \pm 1$$

The radial wave functions R for these states are shown in Figure 3.11. The function for the 2s states is spherically symmetrical with $\psi = 0$ for one value

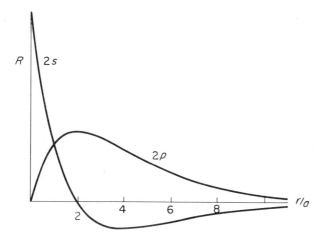

Figure 3.11. 2s wave function and radial part of 2p wave function

of r, i.e. there is a spherical nodal surface. For the 2p states there is no spherical nodal surface but ψ varies with angular direction and each 2p wave function has a nodal plane passing through the nucleus.

The mean *radial* charge densities are shown in Figure 3.12. The maximum value of $R^2 r^2$ for the 2p wave function is at a radius of $4a$.

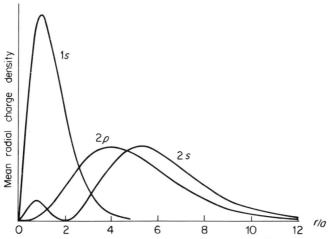

Figure 3.12. Variation of mean electron density with radius for 1s, 2s and 2p states, all to the same scale

The nodal planes for the 2p wave functions can be considered to be perpendicular to a set of rectangular axes, the orientation of which is entirely arbitrary for an isolated atom in the absence of an external electric or magnetic field. A magnetic field has an orientation effect and also affects the energy (see Section 14.2) causing a splitting of the spectral lines. Also when an atom is combined with others, the orientation of the axes is related to the positions of neighbouring atoms (see Chapter 5).

In general, the number of spherical nodes of a wave function is $n - l - 1$ and the number of angular nodes is l, so that the total number of nodal surfaces is $n - 1$ (not counting the nodal surface at infinity).

The photons emitted by hydrogen due to electrons moving from a higher energy state to a lower one will therefore have energies given by

$$\Delta E = \frac{me^4}{8\varepsilon_0^2 h^2}\left(\frac{1}{n_1^2} - \frac{1}{n_2^2}\right)$$

By comparison with equation (2.1) and because $1/\lambda = v/c = \Delta E/hc$ the Rydberg constant is given by

$$R_H = \frac{me^4}{8\varepsilon_0^2 ch^3}$$

3.6 Simple harmonic oscillator

Many atomic and molecular vibrations are linear oscillations or can be resolved into components each of which is a linear oscillation. For most cases the variation of force with displacement may be a complicated function. We consider here the case for a simple linear relationship.

Let a particle move in the x direction under the influence of a force $-\alpha x$, i.e. a force proportional to its distance from the origin and directed towards it. Then the potential energy of the particle when at x will be

$$V = \tfrac{1}{2}\alpha x^2$$

According to classical mechanics, its motion satisfies the equation

$$m\frac{d^2 x}{dt^2} + \alpha x = 0$$

the solution of which is

$$x = x_0 \cos(2\pi v t + \theta)$$

where the frequency v is given by

$$v = \frac{1}{2\pi}\sqrt{\frac{\alpha}{m}} \qquad (3.17)$$

The amplitude x_0 is arbitrary and so the energy E may take any value given by

$$E = \frac{1}{2}m\left(\frac{dx}{dt}\right)^2 + \frac{1}{2}\alpha x^2$$

$$= \tfrac{1}{2}\alpha x_0^2$$

When treated by wave mechanics, Schrödinger's equation for this example becomes

$$\frac{d^2\psi}{dx^2} + \frac{8\pi^2 m}{h^2}(E - \tfrac{1}{2}\alpha x^2)\psi = 0$$

with the boundary conditions that $\psi = 0$ for $x = \pm\infty$. The method of solving this equation is given in Appendix 4 from which it is seen that only certain solutions are possible, and the corresponding energies are given by

$$E = hv(n + \tfrac{1}{2})$$

where v is the frequency of oscillation as defined in equation (3.17) and n can be zero or any positive integral value.

Thus the wave-mechanical treatment leads to the quantization of the possible energy values of a simple harmonic oscillator. Any changes of energy must be in multiples of hv as postulated by Planck (see Section 2.2), but the lowest energy state is not zero.

The normalized wave equations for the quantum numbers $n = 0, 1, 2$ are

$$\psi_0 = \left(\frac{\beta}{\pi}\right)^{1/4} e^{-\beta x^2/2}$$

$$\psi_1 = \left(\frac{\beta}{\pi}\right)^{1/4} (2\beta)^{1/2} x\, e^{-\beta x^2/2}$$

$$\psi_2 = \left(\frac{\beta}{\pi}\right)^{1/4} 2^{-1/2}(2\beta x^2 - 1)\, e^{-\beta x^2/2}$$

where β is given by

$$\beta^2 = 4\pi^2 m\alpha/h^2$$

These three wave functions are shown graphically in Figure 3.13.

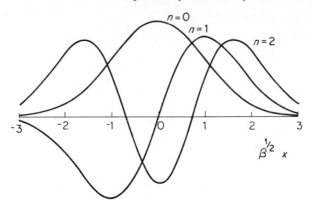

Figure 3.13. Wave functions for simple harmonic oscillator corresponding to quantum numbers $n = 0, 1$ and 2

Questions for Chapter 3

1. A beam of electrons with kinetic energies of 100 eV meets a potential jump of -50 V. Calculate the ratio of the transmitted current to the incident current.

2. There are N electrons in a box of length L and cross-sectional area A. These electrons are uniformly distributed at all times and move only in the direction parallel to the side of length L. The density of electrons is low enough so that any interaction may be neglected and all the electrons are assumed to have an identical energy E. Regarding the box as a well of infinite potential depth derive a quantum wave equation for these moving electrons with an appropriate value to $|\psi|^2$. State any necessary relation between E and L. Find the current impinging on one end wall.

    Show how to obtain a rough *estimate* of the current that flows out of the box if the end wall is now regarded as a potential barrier of height $V(>E)$ and thickness a. [EST]

3. Consider the one-dimensional problem of an electron which is constrained to move along a straight line under conditions such that its potential energy V is related to its distance x from a fixed point O by

$$V = 0 \quad \text{for} \quad -d < x < d,$$
$$V = V_0 \quad \text{for} \quad x < -d \quad \text{or} \quad x > d$$

Beginning with the Schrödinger equation

$$\frac{d^2\psi}{dx^2} + \frac{8\pi^2 m}{h^2}(E - V)\psi = 0$$

Wave Mechanics Applied to Single Electrons

derive expressions from which the permitted energy levels for $E < V_0$ may be calculated.

Calculate the value of the energy of the lowest level, if

$$V_0 = 1 \text{ electron-volt}$$
$$d = 10^{-6} \text{ m}$$
[EST]

4. Derive an expression for the number of electrons with energy between E and $E + dE$ per unit volume in a solid, assuming a uniform internal potential distribution. Use the concept of momentum space and the relationship

$$p = \frac{h}{L}n$$

where p = momentum, h = Planck's constant, L = length of one side of a cube of the solid and n = the quantum number.

Calculate the number of energy states with energies up to 1 eV in 10^{-6} m^3 of material. [EST]

5. Using the time independent general Schrödinger wave equation

$$\nabla^2 \psi + \frac{8\pi^2 m}{h^2}(E - V)\psi = 0$$

determine the numerical values for the most probable distance of the electron from the nucleus of the hydrogen atom and also the ionization energy. Assume that $V = -e^2/4\pi\varepsilon_0 r$ and that a spherically symmetrical solution only is required, for which

$$\nabla^2 = \frac{\partial^2}{\partial r^2} + \frac{2}{r}\frac{\partial}{\partial r}$$

6. The potential energy of a harmonic oscillator can be expressed as

$$V(x) = \tfrac{1}{2}m\omega_0^2 x^2$$

By substitution in Schrödinger's equation in the one-dimensional form, find the possible energy values and the normalized wave equation for the state of lowest energy.

CHAPTER FOUR

The Many-electron Atom

4.1 Wave equation for two or more particles

The behaviour of a single electron moving in the vicinity of a proton, i.e. the case of a hydrogen atom, has been treated by wave mechanics in Section 3.5. For a single electron moving around a nucleus with a greater charge $+Ze$, the wave-mechanical treatment is the same, except that the potential energy at any point will be increased by a factor Z and the resulting quantized energies will be increased by a factor Z^2. Thus the energy of a particular state of an electron in a singly-ionized helium atom would be four times the energy for the corresponding state in a hydrogen atom.

When more than one electron is present in a system, such as an atom, the potential energy of one electron is affected by the presence of the others, i.e. there is some interaction which needs to be taken into account when considering Schrödinger's equation. The motion of two or more particles which form such a system can be represented by a single wave function ψ which is the product of the wave functions $\psi_1, \psi_2, \ldots, \psi_n$ for the separate particles.

Each of the wave functions $\psi_1, \psi_2, \ldots, \psi_n$ will be a function of three independent variables, so that ψ will be a function of $3n$ variables, which may be completely independent.

4.2 The determination of energy levels

Even with only two electrons in the field of a single nucleus where the potential energy of one electron is

$$-\frac{Ze^2}{4\pi\varepsilon_0 r_1} + \frac{e^2}{4\pi\varepsilon_0 r_{12}}$$

r_1 and r_{12} being the distances of that electron from the nucleus and from the other electron, respectively, the problem cannot be solved by exact analytical methods. An approximate numerical method of calculation has to be adopted.

The Many-electron Atom

One method which has been used by Hartree* (the self-consistent field approximation) is easy to understand in principle. Hartree considered that the electrons moving in the field of a nucleus of charge $+Ze$ produce a space charge which is approximately spherically symmetrical. The total field acting on an electron is then due to the attraction of the nucleus and to the repulsion of the negative space charge, i.e. the other electrons partially shield the nucleus.

At the beginning of a calculation, a spherically symmetrical distribution of space charge density due to all but one electron is assumed and the wave function for that one electron in the resultant potential field is calculated. The procedure is repeated for each electron in turn and the charge densities computed are compared with those assumed initially. The degree of agreement is a measure of the reasonableness of the initial choice. The set of calculated values is a better approximation to the correct solution than the initial guess and is used for a repeat calculation. The procedure is repeated until a sufficiently accurate solution is reached.

While s states and completely full p and d states give spherical symmetry of charge (see Section 4.5), this is not so for single p and d states. However, the assumption of spherical symmetry of charge makes for considerable simplification of the calculations and does not lead to gross errors.

It was found that, as in hydrogen, the number of quantized states, which are characterized by similar sets of quantum numbers n, l and m_l, was infinite. Different combinations of states for the electrons could be chosen in performing the calculation and would give different energy values for the atom, corresponding to different excited states of the atom.

A principle which is universally applicable is that the most stable configuration of any system is that for which the total energy is a minimum. Thus, for the ground state (or state of lowest energy of a multi-electron atom) it would seem that all the electrons should be in the state of lowest energy, i.e. the 1s state. However, evidence from spectra shows that this is not so.

4.3 Pauli's exclusion principle and electron spin

The necessary restriction is given by Pauli's exclusion principle, which had to be assumed to account for the observed atomic structure. The principle states that no two electrons can exist with the same wave function or the same set of quantum numbers, including the spin quantum number, which will now be discussed.

To explain certain spectral observations, such as the doublet structure of the spectral lines of alkali metals (a well-known example is the yellow D-line in the sodium spectrum, which can be resolved into two lines with wavelengths of 5890 and 5896 Å), it was necessary to assume that an electron is spinning such

---

* D. R. Hartree, *Proc. Camb. Phil. Soc.*, **24**, 89 and 111, 1927–8.

that its angular momentum about the axis of spin is $\frac{1}{2}(h/2\pi)$ and that there is an associated magnetic moment $eh/4\pi mc$. Further reference to the magnetic moment will be made in Chapter 15. The spin axis can be parallel or anti-parallel to some fixed direction, the two possibilities being denoted by a fourth quantum number m_s which can take either of the values $\pm\frac{1}{2}$. The value of m_s has in some cases a small effect upon the energy levels giving, for example, an energy difference of 0·002 eV between two sub-levels for the $3p$ level in sodium. This accounts for the doublet structure mentioned above.

Dirac, in developing Schrödinger's equation for the electron and taking account of relativistic effects, found that the fourth quantum number appears as a natural result in the same manner as the three quantum numbers appear in Schrödinger's treatment.

With spin included, Pauli's exclusion principle means that there can be two electrons in each s state, corresponding to

$$l = 0, \qquad m_l = 0, \qquad m_s = \pm\tfrac{1}{2}$$

For a p state ($l = 1$), m_l equals -1, 0 or $+1$ and m_s takes two values for each value of m_l giving six possible combinations of quantum numbers. For a d state there will be ten electrons, and fourteen for an f state.

The restriction of Pauli's exclusion principle that no two electrons can have identical wave functions refers to exact identity of the wave functions in all respects. Thus a wave function of an electron in one state in one atom and a wave function with the same set of quantum numbers in another atom are not to be regarded as identical. For the ground state of any atom, the electrons will be in the lowest energy levels consistent with the exclusion principle.

4.4 Splitting of energy levels

In the case of the hydrogen atom, all wave functions for one value of n were degenerate. This is not so for other atoms.

We have seen from Figure 3.12 that the radial charge density or $|\psi|^2$ distributions for a $2s$ and a $2p$ electron are different, the $2s$ electron having a higher probability of being in the region $r < a$, i.e. close to the nucleus, though also a higher probability of being in the region $r > 4\cdot5a$. An electron in either of such states in a multi-electron atom is moving not only in the field of the nucleus but also in the field of the other electrons. The wave function for the $2s$ state penetrates closer to the nucleus and so the electron spends some time in a region where there is less shielding effect due to the $1s$ electrons and its potential energy is lower. The energy of the state is consequently lowered so that states of the same n value, but of different l value, are no longer degenerate.

The energy still depends primarily on the value of n, but it increases as one goes from s to p to d electrons for any one value of n.

The Many-electron Atom

For wave functions with the same value of n but different values of l, the radii at which the radial charge density is a maximum and which may be considered as the radii of the orbits do not differ greatly, while the radii for different values of n are quite distinct. Hence it is convenient to think of the electrons as being in concentric shells, one for each n value.

4.5 Electron structure of the elements

The two electrons of helium will, in the ground state, be in $1s$ states. The lithium atom has three electrons of which two will occupy $1s$ states and one will be in a $2s$ state. Beryllium ($Z = 4$) will have two electrons in each of these states, boron ($Z = 5$) will in addition have an electron in a $2p$ state and so on.

When all occupied shells are full, or contain eight or eighteen electrons, the electron charge distribution is completely symmetrical. The s-type wave functions are always spherically symmetrical. The p and d types are not, but a full set of p or d wave functions for one energy level does have complete symmetry, as is shown in Appendix 5. Such a structure is extremely stable and will not usually react chemically. Thus neon ($Z = 10$) with two $1s$, two $2s$ and six $2p$ electrons is an inert gas.

Also, as discussed more fully in Section 5.4, atoms with loosely bound electrons outside closed shells tend to lose these electrons easily to form positive ions. Atoms which are one or two electrons short of the eight necessary to form a symmetrical charge distribution readily acquire electrons to form stable negative ions. These are known as electropositive and electronegative elements, respectively. Also both types may share outer electrons with other atoms to form covalent bonds. In the elements of the second period (see Figure 1.1), with increase of Z, the $n = 2$ electrons become progressively more tightly bound, because the nuclear charge is increasing, while the $1s$ shell gives a constant shielding charge of $-2e$ in every case.

The third period, from sodium ($Z = 11$) to argon ($Z = 20$), builds up in a similar way, adding firstly two electrons in the $3s$ state and then six in the $3p$ state. Each element corresponds closely to one in the previous period which has the same number of outer electrons around a core of completed shells.

Due to the splitting of the energy levels of each shell, for elements of higher Z, the $4s$ state has less energy than the $3d$ state, so that the atomic structures of potassium and calcium have one and two $4s$ electrons, respectively, with eight electrons in the $n = 3$ shell. The $4s$ electrons are loosely bound to the atoms so that the elements resemble closely sodium and magnesium, respectively, in their chemical behaviour.

The next ten elements in increasing order of atomic number will be formed by filling the $3d$ shell, there being two $4s$ electrons in each case except for chromium and copper, which have only one $4s$ electron each. These ten elements are the transition elements referred to in Section 1.4. The radius of maximum charge density of these $3d$ electrons is much less than that of the $4s$ electrons, so that the $3d$ electrons are better shielded from outside influences. They take less part in chemical bonding than the $4s$ electrons and so all the transition elements show rather similar chemical properties. The $3d$ electrons play an important role in the magnetic behaviour of the transition elements, as will be explained in Chapter 15.

Beyond zinc ($Z = 30$), the $4p$ states are progressively filled for the next six elements up to the inert gas krypton ($Z = 36$).

The eighteen elements of the fifth period follow a pattern similar to that of the fourth, but there is an added complication in the sixth period. In building up successive elements, the first two electrons go into $6s$ states, then one into a $5d$ state, after which the as yet vacant $4f$ states fill up. These electrons are screened by the $n = 5$ and $n = 6$ shells and so exert very little influence on chemical behaviour. These fifteen elements, from lanthanum ($Z = 57$) to lutetium ($Z = 71$), are the rare earth elements. After lutetium, the period completes in the same manner as the previous period.

The seventh period, which is incomplete, appears to follow the same pattern as the sixth.

The electron structures of all the elements are given in Appendix 6.

4.6 X-ray spectra

Using Hartree's method (or any other) the approximate energy levels of any electron in an atom can be calculated. For a heavy atom, the energies of the inner electrons are of a high order of magnitude (8960 eV for the $1s$ electrons in copper). If an electron travelling with kinetic energy of at least that value collides with one of these inner electrons, it can remove the electron from the atom. The vacancy thus created is filled by another electron jumping from an outer orbit with a consequent evolution of energy. This is given out as a photon, the wavelength of which will be of the order of magnitude 0·1–10 Å. This intensely energetic radiation is known as X-rays.

These X-rays can be generated by bombarding a target of the material with a beam of electrons which have sufficient kinetic energy to eject some electrons from the inner shells.

Each element has its characteristic X-ray spectrum, the lines of which are denoted partially by a letter indicating the shell into which the electron jumps. The letters K, L, M, N, O, P and Q refer to the shells for which $n = 1, 2, 3, \ldots, 7$, respectively. The energy levels and the electron jumps which cause the various lines of the characteristic X-ray spectrum of copper are shown in Figure 4.1. The electrons in the energizing beam need energies of at least 8·05 keV to excite the K lines of the spectrum.

The Many-electron Atom

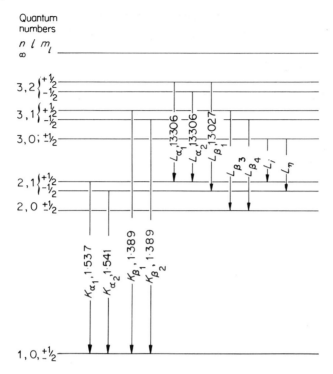

Figure 4.1. Electron energy levels in copper atom and wavelengths (in ångstroms) of X-ray emission spectrum

The energy levels of the inner electrons in a many-electron atom are determined mainly by the nucleus so that electrons in shells of the same and lower quantum number are little affected by the outer electrons.

Hence there is a simple relationship between the frequencies for corresponding lines in the spectra of different elements and atomic numbers.

This is embodied in Moseley's law which can be given mathematically as

$$v = R_H c(Z - a)^2 \left(\frac{1}{n_1^2} - \frac{1}{n_2^2} \right) \tag{4.1}$$

where R_H has the meaning given in Section 2.4, Z is the atomic number, a is a *screening constant* and n_1 and n_2 are the principal quantum numbers of the final and initial levels of the electron.

For the K lines, i.e. $n_1 = 1$, a has the value of approximately one for all elements and for the L lines, i.e. $n_1 = 2$, a is approximately 7·4.

Questions for Chapter 4

1. Write the time-independent Schrödinger's equation for a helium atom in terms of the distances, r_1 and r_2, of the two electrons from the nucleus and r_{12} the distance between the electrons.

2. Discuss the Pauli exclusion principle in relation to the occupancy of permitted energy states by the electrons of an atom, with special reference to scandium.
   [EST]

3. Calculate from equation (4.1) the wavelengths of the K_α, K_β and L_α lines of the X-ray spectrum of copper and compare with the values shown on Figure 4.1.

Chapter Five

Interatomic Bonding

5.1 Overlapping wave functions

At large distances from one another, atoms are completely independent, or, described in wave-mechanical terms, the wave functions do not overlap. When two atoms are brought nearer together, there is some overlap and hence a modification of the potential field in which the electrons move. This, in turn, causes a modification of the wave functions and a change in the energy levels of the electrons. If the total energy of the system is increased by this process, then energy has to be supplied to bring the atoms together, so that there is a repulsive force between them. Conversely, if the total energy decreases as the atoms are brought together, then there is an attractive force.

In almost all cases, the wave-mechanical equations cannot be solved by exact analysis and, as was the case for the multi-electron atoms, numerical methods must be used. In this chapter, only a qualitative treatment of interatomic forces will be given.

As a preliminary to the hydrogen molecule, consider the case of two electrons free to move in one dimension, each electron initially being in a rectangular-sided potential well of width a and depth V_0, and the wells separated by a barrier of width b as shown in Figure 5.1(a).

If $b \gg a$, then the electrons will have real stationary states which will resemble closely the stationary states for an electron in a single potential well of width a, which were discussed in Section 3.3. The ground states or states of lowest energy will be discussed here. Two or more electrons in a system can be described by a single wave function, as was stated in Section 4.1. In this case, the complete wave function can take either of the forms shown in Figure 5.1(a), the upper of which is known as symmetrical and the lower as antisymmetrical. Because the probability of finding an electron at a certain point is proportional to ψ^2, both of these forms will give the same result (except at the mid-point of the barrier, where ψ is small if $b \gg a$) and correspond to the same energy value.

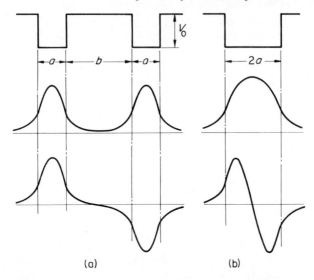

Figure 5.1. Wave function for rectangular potential boxes: (a) when separated by distance b, (b) when brought together

In the limiting case, when b becomes zero, the wave functions will be those for a single potential well of width $2a$ with no node and one node, respectively, as shown in Figure 5.1(b). If the barrier is high compared with the energy level, then it is already known from Section 3.3 that these wave functions correspond to the $n = 1$ and $n = 2$ quantum states, the energy levels of which are approximately in the ratio 4:1, so that the energy of the antisymmetrical state is approximately four times that of the symmetrical state.

5.2 Hydrogen molecule—covalent bonding

The case just considered may be regarded as a one-dimensional simplification of the hydrogen molecule, which consists of two protons and two electrons. The protons are much more massive than the electrons, and to a first approximation may be considered to remain at rest. An electron is attracted by both the protons so that its potential energy is the sum of two terms

$$-\frac{e^2}{4\pi\varepsilon_0 r_a} - \frac{e^2}{4\pi\varepsilon_0 r_b}$$

where r_a and r_b are the distances of the electron from the two protons. The variation of potential along the line joining the protons is as shown in Figure 5.2, there being two wells of infinite depth.

If the distance r_{ab} between the protons is varied, then we shall have a problem similar to the previous case with the energy level corresponding to the ground

Interatomic Bonding

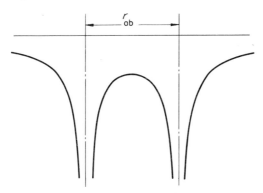

Figure 5.2. Potential due to two protons along line joining them

state in a single hydrogen atom split into two and the two levels becoming more widely separated as r_{ab} is decreased.

Now an electron in a 1s state around a single hydrogen nucleus would have a wave function which varies as $e^{-r/a}$ (see Section 3.5). So in this case, assuming that the form of the wave function is not greatly changed, the wave function would be $e^{-r_a/a}$ around proton A and $e^{-r_b/a}$ around proton B. But the complete wave function may be either symmetric or antisymmetric, so that the wave function for the combined case will be

$$\psi \simeq e^{-r_a/a} \pm e^{-r_b/a}$$

and the shapes of the function along the line of centres will be as shown in Figure 5.3.

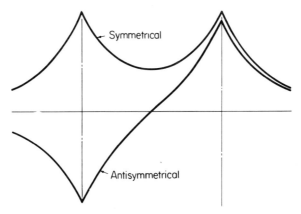

Figure 5.3. Symmetrical and antisymmetrical wave functions for potential field of Figure 5.2

The main difference lies in the region mid-way between the protons or nuclei. The symmetrical wave function is about twice as large in this region as it is at a similar distance from the nucleus in the opposite direction and hence gives an electron charge density which is four times as large. On the other hand, the antisymmetrical function gives no charge density at the mid-point. Therefore the symmetrical function corresponds to an excess charge distribution between the nuclei and an antisymmetrical function to a charge deficiency.

The potential in this mid-region is lower than it would be if the nuclei were more widely separated and thus the symmetrical wave function results in a lower potential energy than for infinite separation. On the other hand, with the antisymmetrical wave function, the electronic charge is forced away from the position of lowest potential energy and consequently the energy for this state is higher.

When the nuclei are so far apart as to be virtually independent, the energy level for an electron in its ground state, i.e. $1s$, is $-R_H hc$ (see Section 2.4). Now when $r_{ab} \rightarrow 0$, the two nuclei virtually become a helium nucleus with $Z = 2$. If only one electron is present, i.e. an He^+ ion, the symmetrical state then corresponds to the $1s$ state of helium with energy $-4R_H hc$. The antisymmetrical state has a nodal plane perpendicular to the line joining the nuclei and when $r_{ab} \rightarrow 0$, this is a nodal plane through the nucleus—which corresponds to a $2p$ wave function with energy $-R_H hc$. Hence the energy of the symmetrical wave function varies from $-R_H hc$ for $r_{ab} = \infty$ to $-4R_H hc$ for $r_{ab} = 0$ and the energy for the antisymmetrical wave function is $-R_H hc$ at both ends of the range. An exact solution would show that for the latter case the energy falls to a minimum at an intermediate value of r_{ab} as may be seen in Figure 5.4 which shows the energy variation for both states.

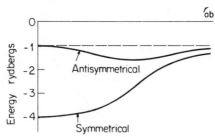

Figure 5.4. Variation of total electron energy with distance between two protons.
1 Rydberg = $R_H hc$

When two electrons are present, as is normal in the hydrogen molecule, the potential is modified due to the interaction of the two electrons. Qualitatively, each wave function will be of the same form as for the single electron case and

Interatomic Bonding

be either symmetric or antisymmetric. The lowest energy state will be that for which both are symmetric and this will be possible only if the electrons have opposite spin.

Both nuclei being positive, there will be a repulsive force between them and associated with this force is a potential energy

$$+\frac{e^2}{4\pi\varepsilon_0 r_{ab}}$$

which must be added to the electron energies to give the total energy of the molecule. The variation of energy with separation of the nuclei will then be as shown in Figure 5.5.

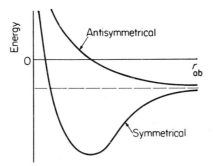

Figure 5.5. Variation of total energy of two hydrogen atoms with distance between them

When both electrons are in the symmetrical state, the energy at first falls as the nuclear distance decreases and then increases to infinity for $r_{ab} = 0$. With one electron in the antisymmetrical and one in the symmetrical state, or with both in the antisymmetrical state, the energy increases with decrease of r_{ab} for all values of r_{ab}.

The stable form of the molecule will be that for which the total energy is a minimum, i.e. the internuclear distance will be given by the minimum of the lower curve in Figure 5.5.

The electrostatic charge density is greatest in the region between the two nuclei and it is the electrostatic attraction of this charge for each nucleus that provides the binding. This is one example of covalent bonding, more examples being given later. It is commonly thought of as a sharing of two electrons, usually one from each of the atoms taking part.

It has been assumed in this discussion that the nuclei are rigidly held at the distance r_{ab}. In practice, the complete wave equation should embrace the nuclei also and allow for their movements. It is found that the equation can be separated into a part which deals with the electrons and another which deals with the nuclei. The latter part is the equation for two concentrated masses separated by a distance r (Figure 5.6). Any motion of the masses can be considered as a transla-

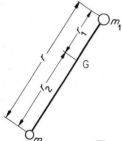

Figure 5.6. Rotating diatomic molecule. G is centre of gravity

tional motion of the centre of gravity G combined with movement of the masses relative to G. The wave equation can be separated into two parts, each dealing with one of these motions. The first, dealing with translational motion, has solutions of the form given in equation (2.15), i.e. a progressive wave for which the total energy is not quantized but may have any value. The second part is similar in form to that for an electron in a hydrogen atom and may be solved in a similar manner by separating the variables for radial and angular motions.

The angular motion will be rotation about axes through the centre of gravity of the system. The radial motion will be a vibration or oscillation of the two nuclei along their line of centres. The energies for both these forms of motion are quantized and so may give rise to specific spectral lines. The vibrational energies are dependent upon the manner in which the potential energy varies with r_{ab}. Solutions for a particular case were discussed in Section 3.6. The rotational energy can take the values

$$E = \frac{j(j+1)h^2}{8\pi^2 I} \tag{5.1}$$

where I is the moment of inertia of the molecule about an axis through G and j is a quantum number which can be zero or a positive integer. The interpretation of molecular spectra will give the possible changes of energy for rotational and vibrational modes and so be of great use in the study of molecular shape and interatomic bonding within the molecule.

5.3 Force between inert gas molecules

When two helium atoms approach one another, there will be possible symmetrical and antisymmetrical forms of the atomic 1s wave function. By Pauli's

Interatomic Bonding   69

exclusion principle, each of these states can take two electrons, and, there being four electrons in all, two will go into the symmetrical and two into the antisymmetrical state. The decrease of energy for the symmetrical state is not sufficient to counterbalance the increase of energy due to the antisymmetrical state and due to the nuclear repulsion. Hence the total energy is found to increase with decreasing internuclear spacing at all spacings so that the helium atoms repel one another and will not form a stable molecule.

In general, any atoms with closed shells of electrons, i.e. complete $1s$ shells, $s + p$ shells of eight electrons, $s + p + d$ shells of eighteen electrons, etc., whether they be inert gases like helium or neon, or ions like Na^+ or Cl^-, will repel one another, when sufficiently close for the wave functions to overlap.

5.4 Ionic or polar bonding

It has already been stated in Section 4.5 that an alkali metal atom will readily lose the single electron in the outer shell to become a positive ion. Energy has to be supplied, the quantity required per atom to do this, expressed in electron-volts, has the following values:

$$\text{Li } 5.4 \quad \text{Na } 5.2 \quad \text{K } 4.3 \quad \text{Rb } 4.2$$

In the presence of electrons, a halogen molecule will become two negative ions with the release of some energy

$$Cl_2 + 2\,e^- \rightarrow 2\,Cl^-$$

The energy released per atom (in electron-volts) for the various halogens is:

$$\text{F } 4.1 \quad \text{Cl } 3.7 \quad \text{Br } 3.5 \quad \text{I } 3.2$$

Hence for the reaction

$$Na + \tfrac{1}{2}Cl_2 \rightarrow Na^+ + Cl^-$$

energy equal to 1·5 eV must be supplied to form each ion pair. However, when the ions are formed, there will be an electrostatic interaction between them, the energy of which will be $-14.4/r$ eV, where r is the interionic distance expressed in ångströms. Therefore, if the ions when formed are less than about 10 Å apart, the electrostatic energy will be more than sufficient to permit the ionization and for all smaller distances the ion pair will be stable.

As r decreases, the total energy of the system will decrease until the ions become so close that the wave functions of the complete shell overlap and give a repulsive force. This is a short-ranged force, but at small values of r, it increases more rapidly than the electrostatic attraction so that the energy curve has a minimum corresponding to a stable interatomic distance.

5.5 Covalent and ionic bonds

The reaction whereby sodium and chlorine form an ion pair can be presented to show the electron structure of the outermost electron shell of each atom as follows:

$$\text{Na} + \text{Cl} \longrightarrow [\text{Na}]^+ [\text{Cl}]^-$$

In calcium chloride, the $+2e$ charge on the divalent calcium ion is balanced by the charges on two monovalent chlorine ions:

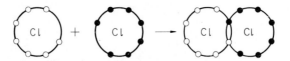

We have seen that the covalent bond between two atoms is a sharing of a pair of electrons with a common level. When the two atoms are of the same kind, the system will be symmetrical, as, for example, when two chlorine atoms are combined in a chlorine molecule.

$$\text{Cl} + \text{Cl} \longrightarrow \text{Cl}\,\text{Cl}$$

Although at times both electrons may be around one nucleus, the average charge distribution will be symmetrical. This form of bonding is favoured by hydrogen and by atoms which have four or more electrons in their outer shell. As many bonds are formed as will bring the number of electrons up to a stable, inert-gas structure.

Between the alkali metals (Li, Na, etc.) at one end of the periodic table and the halogens (F, Cl, etc.) at the other end the elements are less definite in character than these extreme groups. The tendency is for the formation of covalent rather than ionic bonds.

In general, atoms forming positive ions are seldom able to lose more than three electrons, or those forming negative ions to gain more than two. This may

Interatomic Bonding

be easily understood; for example although aluminium loses three electrons to form Al^{3+}, this tendency is strongly opposed by the attraction of the triply charged ion for the electrons.

Fully ionic and fully covalent bonding are completely different, but even so only represent extremes, since many bonds are partly ionic and partly covalent in character. An example is found in the compound between hydrogen and chlorine. These elements each form diatomic molecules with covalent bonding. In the HCl molecule, the bonding is covalent, but the electrons are concentrated more around the chlorine atom than the hydrogen atom.

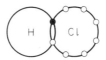

This means that there is a resultant small negative charge at the chlorine end and a small positive charge at the hydrogen end of the molecule. This constitutes an electric dipole—the electrical equivalent of a bar magnet:

$$+ \text{———} -$$

When dissolved in water, the bonding of most of the hydrogen chloride molecules changes to ionic, dissociation to H^+ and Cl^- ions occurring.

A symmetrical molecule, like hydrogen, is a temporary dipole at those times when the electrons in their motion are not symmetrically distributed, as discussed in Section 5.8.

5.6 Directional properties of covalent bonds

In discussing the covalent bonds present in molecules, it is usual to refer to the wave functions of any particular quantum state of electrons as the *electron orbitals*.

The binding energy of a covalent bond is greatest when the electron orbitals of the electrons comprising the bond overlap as much as possible. When an orbital is concentrated in one direction, then the bond formed by it will tend to lie in that direction and also will be stronger because the overlap can be greater.

Now an *s* function is spherically symmetrical, while *p* wave functions have a sort of dumbell shape. The three types of $2p$ wave functions we have considered (Section 3.5) correspond to the nodal plane being perpendicular to the x, y and z axes, respectively, of Figure 3.8 and will be denoted in this chapter by p_x, p_y and p_z, respectively (see Figure 5.7). As stated in Section 3.5, in the absence of an external field, the directions of the axes are completely indeterminate.

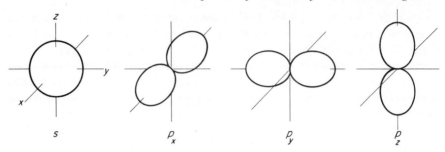

Figure 5.7. Variation of angular function part of wave function with direction for 2s and 2p quantum states

Where the bonding is between s electrons, there will be no preferred direction for the bond. Where p electrons are involved, however, there will be strong directional preference and when more than two atoms are involved in a molecule, there will be geometrical limitations on its shape.

Oxygen has the electron configuration* $(1s)^2(2s)^2(2p)^4$. In any bonding, the 1s or K electrons will not play any part. When oxygen combines with two hydrogen atoms to form a water molecule, the greatest binding energy will be obtained when the hydrogen atoms are so situated that there is the maximum overlapping of electron orbitals.

If one hydrogen atom is located on the x axis, then there will be maximum overlapping between its electron orbital and a 2p oxygen orbital extending along the x axis, i.e. a $2p_x$ orbital. If this orbital contains one electron, it will form a covalent bond with the electron of the hydrogen atom. No further bonding can be carried out by the p_x orbital.

Now if the p_y orbital contains one electron, this can form a bond with a hydrogen atom lying along the y axis. The remaining four electrons of the outer shell of the oxygen atom will be two each in the 2s and $2p_z$ orbitals and can form no more bonds. The expected angle between the bonds of the water molecule would be 90°. It is found by experiment that the angle is approximately 105°. This is due to a repulsion between the hydrogen nuclei. The shape of the molecule is shown in Figure 5.8.

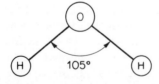

Figure 5.8. Shape of water molecule

The electron charge density is small outside the hydrogen nuclei away from the oxygen atom so that the molecule has resultant positive charges at these

---

* i.e. Two 1s electrons, two 2s electrons and four 2p electrons.

points and a negative charge at the oxygen atom, resulting in a permanent dipole moment.

Nitrogen has an electron configuration $(1s)^2(2s)^2(2p)^3$. An ammonia molecule (NH_3) would presumably be formed by the three hydrogen atoms lying along the x, y and z axes or perhaps slightly shifted from these positions by mutual repulsion. This is found to be so, the shape of the molecule being given in Figure 5.9.

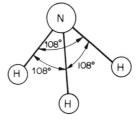

Figure 5.9. Shape of ammonia molecule

A carbon atom has four electrons in its outer shell and is known to combine with four hydrogen atoms in a molecule of methane. The four hydrogen atoms are arranged at the corners of a regular tetrahedron with the carbon atom at the centre as in Figure 5.10. Hence the bonding of each hydrogen atom is apparently

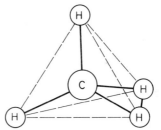

Figure 5.10. Shape of methane molecule

identical. The energies of the $2s$ and $2p$ states in carbon differ very little and this difference can be neglected in comparison with the energies involved in molecular bonding, so that the four electrons can be considered as being degenerate with each other in the molecule but not in the isolated atom. Now any linear combination of a set of degenerate wave functions is also a valid wave function and therefore it is reasonable to take linear combinations which give maximum density in tetrahedral directions. These *hybrid orbitals* are given by

$$\psi_1 = \tfrac{1}{2}(\psi_{2s} + \psi_{2p_x} + \psi_{2p_y} + \psi_{2p_z})$$
$$\psi_2 = \tfrac{1}{2}(\psi_{2s} + \psi_{2p_x} - \psi_{2p_y} - \psi_{2p_z})$$
$$\psi_3 = \tfrac{1}{2}(\psi_{2s} - \psi_{2p_x} + \psi_{2p_y} - \psi_{2p_z})$$
$$\psi_4 = \tfrac{1}{2}(\psi_{2s} - \psi_{2p_x} - \psi_{2p_y} + \psi_{2p_z})$$

where ψ_{2s}, ψ_{2p_x}, ψ_{2p_y} and ψ_{2p_z} are the wave functions for the 2s, $2p_x$, $2p_y$ and $2p_z$ orbitals, respectively. The four hybrid orbitals are exactly equivalent to each other in shape and each is symmetrical about one of the four tetrahedral directions. Thus ψ_1 is symmetrical about the line

$$x = y = z$$

The variation in the electron density with angle is shown in Figure 5.11. The wave function projects in both directions, but one part is much smaller than the other and is unable to overlap effectively with an orbital from another atom. The overlap of these tetrahedral orbitals is greater than would be the case for a p orbital and is the configuration that gives the maximum bond energy.

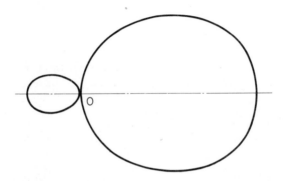

Figure 5.11. Distribution of electron density with direction in hybrid tetrahedral orbital. Radius vector from 0 is proportional to $\int_0^\infty r^2|\psi|^2 \, dr$ along that direction

The larger part of each orbital overlaps with a hydrogen atom in methane and forms the covalent bonds in other organic compounds where carbon is surrounded tetrahedrally by neighbours.

In ethane, C_2H_6, six carbon–hydrogen bonds are formed in a similar way, three for each carbon atom. The fourth orbital of each forms the carbon–carbon bond and maximum overlapping in the bond occurs when the bonding orbital of each carbon atom is directed towards the carbon atom of the other. These two orbitals will be symmetrical about the bond axis so that overlapping is not affected by the rotation of either carbon atom about this bond. Bonds with this characteristic are known as σ bonds.

When two carbon atoms form double bonds, as in ethylene, $CH_2:CH_2$, two orbitals from each atom are involved. The 2s, $2p_x$ and $2p_y$ orbitals can be hybridized to form three trigonal orbitals at 120° to each other in a plane, the $2p_z$ orbital remaining an undisturbed dumbbell shape in the z direction (Figure

Interatomic Bonding

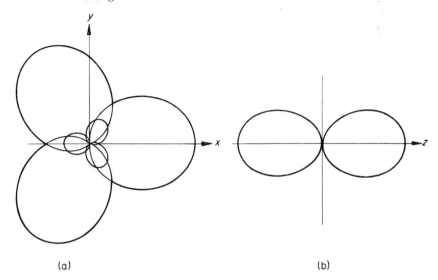

Figure 5.12. Distribution of electron density with direction for hybrid trigonal orbitals and unchanged $2p_z$ orbital; (**a**) in x–y plane, (**b**) perpendicular to x–y plane

5.12). The trigonal hybridized orbital has a shape very similar to that of the tetrahedral orbital and can give almost as much overlap and hence almost as strong bonding. Two of the trigonal orbits of each carbon atom bond with hydrogen and the third of each contributes to a carbon–carbon bond of the σ type. A second carbon–carbon bond results from interaction of the two p_z orbitals and is of a different nature. The concentration of electron charge is greatest in the plane through the carbon nuclei and perpendicular to the plane containing the hydrogen atoms. This is the π-type of bond. There is not much overlap of the orbitals and consequently this bond is much weaker than the σ-type. Also the π bonding prevents free rotation of the carbon atoms about the bond axis so that the ethylene molecule is essentially planar.

In the double bond, four electrons are concentrated in the region between the nuclei and the attraction of the carbon atoms to this region is stronger than is the case in a single bonding. Hence a double bond should be shorter than a single bond—which is found to be so, the carbon–carbon distance in ethylene and ethane being 1·3 and 1·54 Å, respectively.

Also the double bond will be stronger, but since π bonding is weaker than σ bonding, it will not be twice as strong as a single bond. It is, however, easy to rupture the π bond and not the σ bond.

The triple bond in acetylene HC ⫶ CH is formed somewhat similarly by a σ bond and two π bonds between the two carbon atoms, the carbon–carbon distance being still smaller, viz. 1·1 Å.

Inorganic radicals, such as the sulphate ion $(SO_4)^{2-}$, have internal bonding which is most easily regarded as being of the covalent type. In the sulphate ion, the six electrons of the $n = 3$ shell of sulphur, together with the two electrons which the ion has received from metal ions, form eight electrons which resonate between the sulphur and the oxygen atoms. The four oxygen atoms adopt a tetrahedral configuration and the whole group has a resultant electric charge of $-2e$. It can be represented in a two-dimensional diagram as:

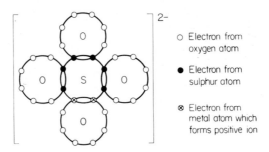

5.7 Metallic bonding

Metals form the major number of the elements. They possess strength, so that there are cohesive forces between atoms but yet are ductile so that the bonding is not such that once broken between two atoms it cannot be repaired. A metal is regarded as being positive ions in a cloud of electrons (the valence electrons) which are not attached to particular atoms. They form a cloud, the attraction of which for the positive ions holds the metal together (in this respect it is like a covalent bond). The electrons have freedom of motion which gives electrical conductivity. The metallic bond is discussed more fully in Chapter 10.

5.8 Intermolecular forces

Molecules are formed when all the atoms in a group are held together by covalent bonds and all the valencies are saturated. Atoms of the inert gases do not combine and so for these substances single atoms may be considered to be molecules. Such molecular compounds can exist as solids below their melting points and as liquids below their boiling points, so that intermolecular forces must exist. However, the energy of the intermolecular forces must be considerably smaller than the interatomic bonding energy within the molecules.

For all sorts of molecules the graph of variation of potential energy with intermolecular spacing will be the difference of the energy-spacing graphs for attraction and repulsion and will have a minimum at the standard intermolecular spacing, r_0, as shown in Figure 5.13. The repulsive forces will be due to the overlap of complete electron shells.

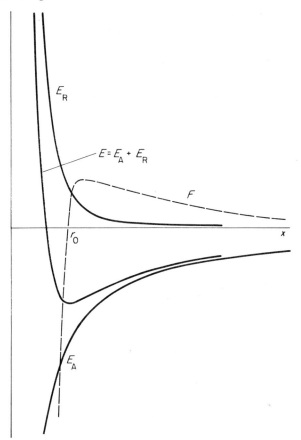

Figure 5.13. Variation of energy of molecule and force on molecule with intermolecular spacing x. Energy is difference of energy of attraction E_A and energy of repulsion E_R. Force $F = dE/dx$

The attractive forces, which are known as van der Waals forces, are due to the forces between the electrical dipoles of the two molecules. Some molecules, like water, have permanent dipole moments and so will attract one another when the positive end of one is directed towards the negative end of the other.

This mode of bonding is referred to as a *hydrogen bond* or *hydrogen bridge*. The hydrogen atom which is covalently bonded to an electronegative atom in the molecule also has an electrostatic attraction, which is weaker than that of an ionic bond, to an electronegative atom of another molecule.

Other molecules, which have no permanent dipole moments may yet have temporary ones. There will be times when the electrons, which are constantly in motion, will not be distributed symmetrically. The temporary dipole moment will cause a surrounding electric field which will tend to induce a dipole moment in any other molecule which lies within the field. The magnitude of the induced dipole will be proportional to the field, which itself, apart from geometrical factors, is proportional to the inducing dipole moment. The sign of the induced dipole is such that there will be an attractive force between the two molecules. Hence over a period of time, the average attractive force is proportional to the mean-square dipole moment.

In general, the larger the number of electrons in a molecule, the larger is the dipole moment that can be formed. As an example, with increase of atomic number in the inert gases, corresponding to more electrons per atom, the higher are the melting and boiling points, which implies a stronger bond.

Questions for Chapter 5

1. Describe the electron structure of the light atoms up to that of atomic number 19. Discuss the formation of ionic and covalent bonding between atoms in terms of the electron structure. [MST]

2. Describe the atomic structure of helium, aluminium and bromine. What types of chemical bond can each form? [MST]

3. Describe the arrangement of electrons in atoms of the alkali metals and the nature of the bond between atoms in crystals of these metals.
   How are some of the physical properties typical of metals explained in terms of this structure? [MST]

4. Describe the various types of bond which exist between atoms in a solid. How does the type of bond influence the mechanical and physical properties of the solid? [MST]

5. The energy of interaction for two ions can be expressed as the energy of attraction k_1/r and the energy of repulsion k_2/r^9, where $k_1 = 27\,\text{eV}\,\text{Å}$, $k_2 = 3 \times 10^4\,\text{eV}\,\text{Å}^9$ and r is the distance between the ion centres.
   Calculate the equilibrium spacing r_0 between the two ions and the distance r_1 for which the net attractive force is a maximum. [P]

6. Assume that the energy of two atoms in each other's field is given by

$$U(r) = -Ar^{-n} + Br^{-m}$$

where r is the distance between centres and A and B are positive constants which determine the strength of the attractive and repulsive forces, respectively. Show that, if the configuration is stable, $m > n$ where m and n are both positive. [MST]

CHAPTER SIX

Aggregations of Atoms–Gases and Liquids

6.1 Introduction

The engineer is usually not concerned with atoms as individuals but with the behaviour or properties of populations of atoms in one of three states of aggregation: gas, liquid and solid. In gases, the spacing of the molecules is large (except at very high pressures) compared to that in liquids and solids where each atom or molecule is in close proximity to its neighbours.

Properties of gases which may be of concern to the electrical engineer include viscosity, which will affect the flow in pumping systems, thermal conductivity and diffusion, as well as ionization, electrical conductivity, etc. The former group can be treated and understood without any knowledge of the internal structure of the molecule or atom and are considered in this chapter in terms of the elementary kinetic theory of gases. Some brief paragraphs on the properties of the liquid state showing their relationships to and differences from gases follow.

6.2 The behaviour of a gas

The more obvious qualitative features of gas behaviour are homogeneity, large compressibility, diffusion of one gas through another and non-settling.

Quantitatively gases at low pressures obey the laws of Boyle and Charles, which are summarized by the equation

$$pV = R_m T$$

where p and V are the pressure and volume of a mass m of the gas at an absolute temperature T, and R_m is a constant appropriate to the mass m. Clearly, R_m is proportional to m, since at the same pressure and temperature V will be proportional to m. A gas which obeys these laws is known as a *perfect gas*. All gases approximate to this behaviour except at high pressures and low temperatures.

There is also Avogadro's hypothesis, which is accurate to the extent that gases obey the simple gas laws. This was deduced originally from chemical evidence and has since been verified in other ways. It states that at the same temperature and pressure equal volumes of all gases contain the same number of molecules.

6.3 The kinetic theory of gases

The behaviour of a perfect gas can be explained by assuming that molecules of a gas are all alike, occupy a volume that is negligible compared with the total volume and are in motion, exerting no force on each other or the container walls except when actually in contact, all collisions being perfectly elastic.

The pressure exerted by a gas is due to the change of momentum of the molecules in collisions with the walls of the container. It can be shown to be given by

$$p = \frac{nm\overline{C^2}}{3V}$$

where n is the number of molecules in a volume V, m is the mass of a molecule and $\overline{C^2}$ is the mean square velocity of the molecules. Hence

$$pV = \tfrac{1}{3} nm\overline{C^2}$$
$$= \tfrac{2}{3} \times \tfrac{1}{2} nm\overline{C^2}$$
$$= \tfrac{2}{3} \times \text{Kinetic energy of translational motion of all the molecules in the gas}$$

If the kinetic energy is proportional to absolute temperature, i.e.

$$\tfrac{2}{3} \times \tfrac{1}{2} nm\overline{C^2} \propto T \quad \text{or} \quad \tfrac{2}{3} \times \tfrac{1}{2} nm\overline{C^2} = RT$$

where R is a constant, then

$$pV = RT$$

If we have equal volumes of two gases at the same temperature and pressure, then

$$\tfrac{2}{3} \times \tfrac{1}{2} n_1 m_1 \overline{C_1^2} = \tfrac{2}{3} \times \tfrac{1}{2} n_2 m_2 \overline{C_2^2}$$

where n_1 and n_2 are the numbers of molecules, and m_1 and m_2 are their masses for the two gases, respectively.

Then if we assume that the average kinetic energies of the molecules of these gases are the same at the same temperature,

$$\tfrac{1}{2} m_1 \overline{C_1^2} = \tfrac{1}{2} m_2 \overline{C_2^2}$$

Aggregations of Atoms—Gases and Liquids

Hence $n_1 = n_2$, or equal volumes of two gases at the same temperature and pressure contain equal numbers of molecules, which is Avogadro's hypothesis.

6.4 Mole

If M is the molecular weight of a substance, then the quantity the mass of which is numerically equal to M is called a *mole*. Thus, a kilogramme mole (kmol) has a mass of M kg.

From the definition of the mole, it follows that one mole of any substance contains the same number of molecules. This number is *Avogadro's number* and its value is

$$N_0 = 6{\cdot}02 \times 10^{26} \text{ kmol}^{-1}$$

If m is taken as the kilogramme mole of any gas, R_m will always have the same value $\bar{R}$:

$$\bar{R} = 8314 \text{ J kmol}^{-1} \text{ K}^{-1}$$

Although the value of $\bar{R}$ is of great importance, it has no fundamental significance, since the size of a kilogramme mole is an arbitrary choice depending on the standard unit of mass and the convention that the mass of the common isotope of carbon be taken as twelve.

Hence it is often convenient to use a more fundamental constant:

$$\frac{\bar{R}}{N_0} = k = \text{Boltzmann's constant}$$

$$= 1{\cdot}38 \times 10^{-23} \text{ J K}^{-1}$$

This may be regarded as the gas constant for a single molecule.

6.5 Maxwellian distribution of velocities

It can be seen from the kinetic theory of gases considered in Section 6.3 that the mean square velocity of gas molecules is given by

$$\overline{C^2} = \frac{3\bar{R}T}{N_0 m}$$

$$= \frac{3kT}{m}$$

The theory does not give any information about the *distribution* of velocities among the molecules, but the distribution can be found by statistical thermodynamic methods. *Maxwell's distribution law*, which is derived in Appendix 7, states that, of the total number of molecules of a gas, the fraction which have velocities lying between c and $(c + \delta c)$ is given by

$$N(c)\,\delta c = 4\pi \left(\frac{m}{2\pi kT}\right)^{3/2} c^2\, e^{-mc^2/2kT} \delta c$$

The curve of the distribution of velocities has the shape shown in Figure 6.1.

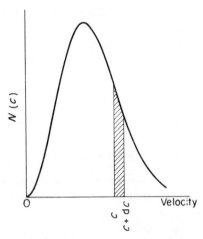

Figure 6.1. Distribution of velocities of molecules of a gas

The shaded area is proportional to the above expression. An increase of temperature alters the shape of the curves as may be seen from the distribution curves for hydrogen in Figure 6.2.

The mean velocity is found to bear a definite relation to the root-mean-square velocity

$$\bar{c} = 0\cdot 921 \sqrt{(\overline{C^2})}$$

The most probable velocity α, i.e. the value of the velocity at the peak of the distribution curve, is given by

$$\alpha = 0\cdot 816 \sqrt{(\overline{C^2})}$$

6.6 Mean free path

The *mean free path* is the average distance that a molecule travels between successive collisions with other molecules.

A simple calculation of its value can be made by assuming all molecules to be rigid spheres and all but one to be at rest. Let

d = diameter of each molecule

n = number of molecules per unit volume

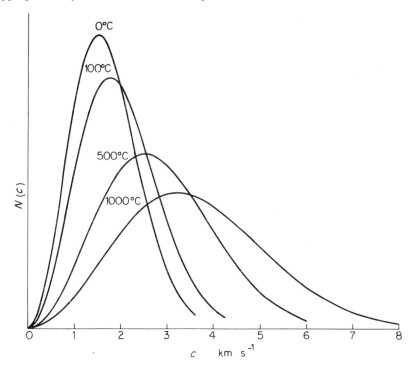

Figure 6.2. Effect of temperature on velocity distribution of hydrogen gas molecules

The moving molecule will collide with any other molecule whose centre lies within a distance d of the path of the centre of the moving molecule. In travelling a distance l, it will collide with all molecules whose centres lie in a cylinder of length l and radius d (Figure 6.3). The number of molecules contained in the volume $\pi d^2 l$ is $\pi d^2 l n$, which is also the number of collisions.

Figure 6.3.

Hence the mean free path λ is given by

$$\lambda = \frac{l}{\pi d^2 l n} = \frac{1}{\pi d^2 n}$$

If account is taken of the motion of all the molecules, with their velocities distributed according to Maxwell's distribution law, the expression is modified to

$$\lambda = \frac{1}{\sqrt{2}\pi d^2 n}$$

The term *diameter* as applied to a gas molecule does not have any precise significance, so that it is not possible to calculate λ from the above expression. The mean free path is obviously of importance in connection with any phenomena associated with the transfer of any property of a gas due to molecular motion. Examples are:

a. *Viscosity*, due to transfer of momentum from one region to another where the gas is moving at a lower speed.
b. *Thermal conductivity*, due to transfer of energy from one region to another at a lower temperature.
c. In a gas containing two or more different kinds of molecules, the relative densities of which vary from point to point, *diffusion* occurs to reduce the variation.

These three processes, which can be treated in a similar manner are known as *transport phenomena*.

6.7 Transport properties

The following approximate method of calculation for these gives the factors upon which each depends, though not the exact value of the numerical constant.

A molecule which describes a free path ST between two successive collisions may be regarded as transporting some property P of magnitude appropriate to position S from S to T. Sooner or later, another molecule will describe the reverse path also transporting some of the property from T to S. If the gas is in a steady state, i.e. uniform in streaming velocity, temperature or composition, then although the magnitude of the property possessed by individual molecules may have a statistical distribution, the average value per molecule does not vary from place to place so that on balance there will be no net transfer of the property.

If, however, the mean value of the property varies from place to place, then the transfers in the two directions will not be equal.

Aggregations of Atoms—Gases and Liquids

Let the mean value of the property P vary uniformly in the z direction such that dP/dz is positive and consider unit area of the plane $z = z_0$.

If the free path ST of one of the molecules crosses this plane (Figure 6.4), then

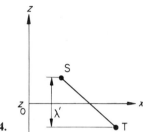

Figure 6.4.

it transfers property P appropriate to position S across the plane. The molecule with a free path TS will transfer in the reverse direction across the plane property P appropriate to position T. If λ' is the projection on the z axis of the free path ST, then the difference in the magnitude of property P between S and T is

$$\lambda' \, dP/dz$$

and this is the net amount transferred across the unit area in the direction of decreasing z due to the passage of one molecule in each direction.

The number of molecules that cross unit area in each direction in unit time is of the order of magnitude of $n\bar{c}$. Hence the rate of transfer of property across unit area of the $z = z_0$ plane would be

$$n\bar{c}\lambda' \frac{dP}{dz}$$

if each free path had the same projection λ' on the z axis. If the free paths are distributed equally in all directions, then the averaged value of λ' for molecules crossing the unit area from one side is $\lambda/2$ where λ is the mean free path, so that the rate of transfer of the property is

$$\frac{1}{2} n\bar{c}\lambda \frac{dP}{dz}$$

A more refined calculation would make the numerical coefficient more nearly equal to $\frac{1}{3}$ instead of $\frac{1}{2}$.*

---

* See Sir James Jeans, *An Introduction to the Kinetic Theory of Gases*, Cambridge University Press, Cambridge, 1940, Chaps. VI, VII and VIII.

6.7.1 Viscosity

In a gas which has a general streaming velocity v_x in the x direction, that varies in the z direction, i.e. there is a velocity gradient dv_x/dz, the property transferred across the $z = z_0$ plane is the average momentum mv_x of the molecules. The rate of transfer across unit area is

$$\frac{1}{3}n\bar{c}\lambda\frac{d(mv_x)}{dz} = \frac{1}{3}\rho\bar{c}\lambda\frac{dv_x}{dz}$$

where ρ is the density of the gas.

If the velocity gradient is to be maintained in the gas, then a shear force must be applied parallel to the $z = z_0$ plane to counterbalance the change of momentum. The shear force or viscous drag τ on unit area of the plane will equal this rate of transfer of momentum. The *coefficient of viscosity* μ is defined as the constant of proportionality between this shear force and the velocity gradient

$$\tau = \mu\, dv_x/dz$$

so that

$$\mu = \frac{1}{3}\rho\bar{c}\lambda$$

6.7.2 Thermal conductivity

If a temperature gradient exists in the gas, then the mean energy $\bar{E}$ of a molecule also varies, the two being related by the specific heat of the gas at constant volume

$$C_v = \frac{1}{m}\frac{d\bar{E}}{dT}$$

The net energy transported across unit area of the $z = z_0$ plane in unit time is

$$\frac{1}{3}n\bar{c}\lambda\frac{d\bar{E}}{dz}$$

But the energy conducted across unit area in unit time is also

$$K\frac{dT}{dz}$$

where K is the *thermal conductivity*, so that

$$K = \tfrac{1}{3}\rho\bar{c}\lambda C_v$$

Therefore the thermal conductivity and coefficient of viscosity are related by

$$K = \mu C_v$$

K can be measured experimentally and hence λ and d calculated. Some approximate values at 15 °C and 1 atmosphere pressure are:

	λ	d
H_2	16.3×10^{-8} m	2.3×10^{-10} m
N_2	8.5	3.15
O_2	9.6	2.96

6.7.3 Gaseous diffusion

Consider a closed vessel containing two different gases A and B and let there be n_A and n_B molecules per unit volume, respectively. The pressure is assumed to be the same throughout so that the total number of molecules per unit volume must be the same everywhere.

Initially let there be a uniform variation of n_A and n_B in the z direction. Then there will be a net flow of each gas along this direction in such a way as to make the mixture homogeneous. This is the process known as *diffusion*.

Since $n_A + n_B$ is constant, then the gradients of their densities must be equal and opposite, i.e.

$$\frac{dn_A}{dz} = -\frac{dn_B}{dz}$$

Also the flows of the two types of molecule across unit area of the $z = z_0$ plane must be equal and are given by

$$I_A = -D\frac{dn_A}{dz}$$

and

$$I_B = -D\frac{dn_B}{dz}$$

where D is the *coefficient of diffusion*.

If the two sorts of molecules have approximately the same size and mass, so that λ and $\bar{c}$ will be approximately the same for them and the differences can be neglected, then the number of one sort transported across unit area of the $z = z_0$ plane is

$$\frac{1}{3}\bar{c}\lambda\frac{dn_A}{dz}$$

so that

$$D = \tfrac{1}{3}\lambda\bar{c}$$

6.8 Variation of λ, μ and K with pressure

Since $\lambda = 1/\sqrt{2\pi d^2 n}$ and because n is proportional to pressure at a given temperature,

$$\lambda \propto \frac{1}{p}$$

Therefore at low pressures λ may be quite large.

For example, in a radio valve, where the pressure is of the order of 10^{-9} atmosphere, the mean free path will be about 100 m. This is much greater than the dimensions of the valve, so that in general a molecule traversing the valve will not collide with any other molecules. (Even at this low pressure, there are still 2.7×10^{16} molecules m^{-3}.)

Because n is proportional to pressure and λ to the reciprocal of pressure, both μ and K will be independent of pressure. This is found to be true at moderate pressures, but not at low ones. Obviously the above expression is no longer valid when the mean free path becomes comparable with the size of the containing vessel.

6.9 Deviations from the gas laws

The pressure–temperature–volume relationship of gases departs from the perfect gas laws at higher pressures and at lower temperatures.

Hydrogen, oxygen and other gases which are liquified only at very low temperatures show negligible departure from perfect gas behaviour at temperatures from 0 °C upwards. Carbon dioxide shows a marked departure below about 50 °C. Andrews conducted experiments on this gas using pressures up to 14 MN m^{-2}, getting the results shown in Figure 6.5. Each curve is an *isothermal* showing the relationship between pressure and volume at constant temperature of a particular mass of gas.

The area enclosed by the broken curve is a region where gas and liquid coexist. For any isothermal, such as that for 21.5 °C the carbon dioxide was gaseous over region pq and liquid over region rs. Liquefaction as the pressure was increased took place for all temperatures up to 31.1 °C above which no liquefaction occurred at any pressure: 31.1 °C is called the *critical temperature* for carbon dioxide, and the isothermal for this temperature is the critical isothermal. The pressure and volume at the maximum point on the broken curve are known as the *critical pressure* and *critical volume*, respectively.

All gases show behaviour of a similar type, but the values of the critical temperatures, pressures and volumes differ widely.

Aggregations of Atoms—Gases and Liquids

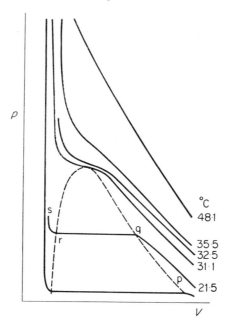

Figure 6.5. Andrews' isothermals for carbon dioxide

6.10 Van der Waals' equation

In deriving the perfect gas laws on the basis of the kinetic theory, two factors were neglected. These were the attractive forces between the molecules and the volume of the molecules. These factors were considered by Van der Waals, who made the first attempt to modify the simple kinetic theory.

The molecules have finite size but at standard temperature and pressure occupy only about 0·1 per cent. of the total volume. When the gas is highly compressed, the volume of the molecules becomes an appreciable part of the total so that a molecule has less far to go between collisions than allowed for in the simple theory and hence will collide with the container walls more frequently giving a higher pressure.

The gas law equation can be written as

$$p(V - b) = RT$$

where b is a correction for the volume of the molecules.

The mutual attraction of the molecules has the same effect as an extra external pressure would have on molecules between which there is no mutual attraction, i.e. it will decrease the volume. An additional *internal pressure* term must therefore be added to the pressure term of the simple theory. The force on a single molecule is proportional to the number of molecules within a distance such that the attractive force due to any of them is appreciable. Also the internal pressure depends upon the number of molecules in a given volume that are subject to the attractive forces. Each of these is proportional to the density, that is to the reciprocal of the specific volume, so that the internal pressure is equal to a/V^2 where a is a constant. The equation of Van der Waals is thus

$$\left(p + \frac{a}{V^2}\right)(V - b) = RT$$

6.11 Comparison of Van der Waals' equation with experiment

Isothermal curves calculated from this equation are plotted in Figure 6.6. At low temperatures each curve has a maximum and a minimum, while at higher temperatures the curves merely have a point of inflexion and get closer to $pV = RT$ with increasing temperature.

One intermediate curve has a point of inflexion with a horizontal tangent at C. This corresponds to the critical isothermal as found by Andrews, and C is the critical point.

Above the critical isothermal the curves agree reasonably with experimental values, but below it Van der Waals' equation gives an S-shaped curve ABDE, whereas experiment gives A$\alpha\beta\gamma$E, with discontinuities at α and γ. Here $\alpha\beta\gamma$ corresponds to the heterogeneous region of two phases in equilibrium.

The portions αB and γD which correspond to a supercooled vapour and a superheated liquid, respectively, can be obtained experimentally. BD represents an unstable region where increase of pressure causes an increase of volume, and could not be realized in practice.

By thermodynamic reasoning it can be shown that the horizontal line $\alpha\beta\gamma$ should be drawn in such a position that areas αBβ and βDγ are equal.

Van der Waals' equation is not completely correct, but is, however, a good general approximation and has the merit of simplicity. Other equations of state have been suggested, which are improvements on that of Van der Waals, but none give complete agreement with experimental results.

6.12 The behaviour of liquids

A liquid, like a gas, is composed of molecules, but with much less free space. The molecules are still freely in motion, having kinetic energy which is dependent upon temperature, but the Van der Waals attractive force has a significant value.

Aggregations of Atoms—Gases and Liquids

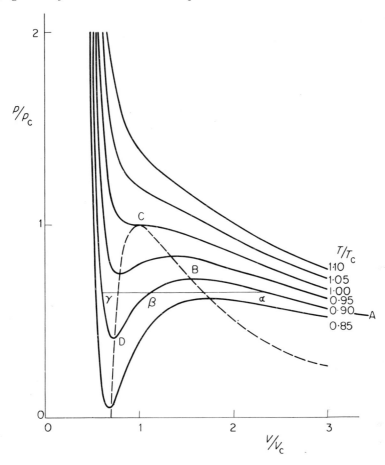

Figure 6.6. Isothermals calculated from Van der Waals' equation. The pressure, volume and temperature are expressed in terms of the values of these variables at the critical point C. In this form, the curves are independent of the actual values of a and b

6.13 Surface tension

Molecules in the bulk of the liquid will experience equal Van der Waals attractive forces in all directions, but those molecules nearer the surface will have an attractive force towards the bulk of the liquid which is not counterbalanced by any force from outside. This force on the surface molecules, which causes the liquid to behave as if enclosed in a skin, is known as the *surface tension* and is measured in terms of the force necessary to separate an element of the surface layer of unit length. In increasing the surface area, work equal to the product of the surface tension and the increase of area has to be done. This is the *surface energy*.

The surface tension decreases with increasing temperature and becomes zero at the critical temperature (where vapour and liquid are indistinguishable). The surface tension of water is about 0·070 and that of alcohol about 0·025 N m^{-1} at room temperature. The surface tension of liquid metals is much higher, for example molten aluminium and iron have values of 0·5 and 1·5 N m^{-1}, respectively.

A drop of mercury resting on a horizontal surface will keep an almost spherical shape, whereas water would spread out to a thin layer because of the much higher surface tension of the metal, although the mercury has a much greater density.

6.14 Vapour pressure

A molecule moving towards the surface will have its velocity reduced as it passes through the region of the unbalanced force. The translational energy of any molecule which has an initial energy equal to or less than the mean value for that temperature will be reduced to zero. Some molecules will, however, have a velocity sufficiently high to escape. They constitute the vapour of the liquid and exert a pressure known as the *vapour pressure* of the liquid. Owing to collisions some of these molecules will later have velocities towards the surface, and when they strike it they will be absorbed into the bulk of the liquid. If evaporation takes place inside a closed container, the number of vapour molecules will at first increase. As the density of vapour molecules increases, so also will the rate at which they return to the liquid. An equilibrium state will be reached when the rates at which molecules leave and reenter the liquid are equal. The vapour is then saturated, and its vapour pressure is called the *saturation vapour pressure*.

When the saturation vapour pressure is equal to or greater than the total external pressure boiling can occur.

6.15 Viscosity

When a fluid is in a state of motion with different layers moving at different speeds, there is shear force between the layers which would bring all the fluid to the same average speed if external forces were not applied. This shear force is described as *viscosity*. In gases, as already discussed in Section 6.7, it is a transport phenomenon and its value is related to the mean free path.

In liquids, the mean free paths are extremely short, so that the molecules will not in general pass from one layer to another. The intermolecular forces are, however, significant, so that each layer tends to drag the adjacent layer with it, thus reducing the relative motion.

Aggregations of Atoms—Gases and Liquids

6.16 Thermal conductivity

The conductivity of heat is also a transfer of molecular kinetic energy. If the temperature of one region in a liquid is raised above that of the surrounding regions, then the mean kinetic energy of the molecules in that region will be greater. The intermolecular forces are considerable even when the molecules are not in direct contact so that there will be a transfer of kinetic energy to the adjoining molecules outside the region. Whereas with viscosity there was a definite transfer of momentum, the average momentum in this case will be zero.

Questions for Chapter 6

1. 0·1015 g of an organic liquid, when vapourized, displaced a quantity of air whose volume was 27.96×10^{-6} m^3 measured at 15 °C and 750 mm of mercury pressure. Assuming Avogadro's hypothesis to hold for the vapour, calculate the molecular weight of the liquid (specific weight of mercury = 133 kN m^{-3}).

2. Calculate the root mean square velocity of a molecule of krypton at 300 °C.
   Given that the mean free path of the krypton molecule at 0 °C and 750 mm of mercury is 9.5×10^{-8} m estimate the molecular diameter. [P]

3. Define the *mean free path* of a molecule of a gas.
   Derive an expression for the mean free path. Using this expression, calculate for argon at 15 °C the pressure at which the mean free path is 0·01 m, assuming that the diameter of the argon atom is 2.88×10^{-10} m. [P]

4. Define *mean velocity, r.m.s. velocity* and *mean free path* of molecules in a gas.
   The velocity of sound in a monatomic gas is given by $(\frac{5}{3}p/\rho)^{1/2}$, where p is the pressure and ρ is the density. Find the ratio of the r.m.s. velocity of the molecule to the velocity of sound in such a gas. [MST]

5. A vacuum tube 30 mm in diameter contains nitrogen at room temperature and at a pressure of 10^{-5} mm Hg. Assuming that the diameter of a nitrogen molecule is 4×10^{-7} mm, show that the molecular paths are usually terminated by collision with the wall of the tube. [MST]

6. What is meant by the mean free path of the molecules of a gas?
   Explain qualitatively, in terms of the mean free path, the properties of viscosity and thermal conductivity of a gas. [P]

7. Explain the assumptions made by Van der Waals in deriving his equation from the simple gas law $pV = RT$, where p, V, R and T have their usual meanings.

State the Van der Waals equation. To what extent can it represent the experimental pressure, volume and temperature relationship of a gas such as carbon dioxide? [P]

8. The critical point of a gas occurs at 10 °C and a pressure of 51 atmospheres, the density then being $2 \cdot 2 \times 10^2$ kg m^{-3}. Assuming Van der Waals' equation to be correct in the region of the critical point, calculate the molecular weight of the gas. (1 atmosphere = 0·76 m of mercury pressure.)

9. Maxwell's law concerning the distribution of the component velocities u, v, w, of the molecules of a gas can be written in the form

$$N(\beta m/\pi)^{3/2}\, e^{-\beta m u^2}\, du\, e^{-\beta m v^2}\, dv\, e^{-\beta m w^2}\, dw$$

where $\beta = 1/2kT$ and N is the total number of molecules in the system. This expression may be assumed to hold at all points in the gas, including those adjacent to the walls of the containing vessel.

Taking this expression as a starting-point:

a. Deduce an expression for the distribution of the total speed of the molecules and use it to determine the most probable speed;

b. Derive an expression for the number of molecules whose x component of velocity lies between u and $u + du$, when there is no limit on the values of the other two components;

c. Taking the molecules to be contained in a cubical box of unit volume, with sides perpendicular to the directions u, v and w, respectively, calculate the number of molecules striking a side in unit time.

Calculate the mean energy of translation of these molecules and explain why the value differs from the mean energy of translation of all molecules in the gas

$$\int_0^\infty e^{-\beta m u^2}\, du = \frac{1}{2}\sqrt{\frac{\pi}{\beta m}}, \qquad \int_0^\infty e^{-\beta m u^2} u\, du = \frac{1}{2\beta m}$$

$$\int_0^\infty e^{-\beta m u^2} u^2\, du = \frac{1}{4}\sqrt{\frac{\pi}{\beta^3 m^3}}, \qquad \int_0^\infty e^{-\beta m u^2} u^3\, du = \frac{1}{2\beta^2 m^2} \qquad \text{[EST]}$$

CHAPTER SEVEN

The Crystalline State

7.1 Introduction

With a few exceptions, most electrical engineers are concerned with the properties of materials which are in the solid state. A true solid is one in which the atoms or molecules occupy relatively fixed positions with respect to each other. These positions might better be described as *centres of oscillation*, the atoms or molecules oscillating about them with amplitudes which increase with increasing temperature.

The fixed positions form a pattern that repeats periodically in three directions. One network of the pattern will continue either until it meets another network which has a different orientation or else until it meets a free surface. In the latter case, regular faces may form naturally, giving the usual elementary idea of a crystal. Where crystals of different orientations meet, the boundary will not necessarily be a plane surface, but is curved or irregular. The sizes of these separate crystals or grains are a function of the past history of the specimen.

Glassy solids and some others which are non-crystalline are termed *amorphous solids* and may be regarded as liquids of very high viscosity. They are considered in Section 8.4.

7.2 Space lattices

Three non-parallel sets of planes, those in each set being parallel and equispaced, will intersect to give a set of points (each being at the intersection of three planes) which form a rectangular pattern called a *space lattice*.

One property of a space lattice is that each *lattice point* has surroundings which are identical to those of any other lattice point. Space lattices are fundamental in the description of crystal structures.

When atoms or clusters of atoms are positioned at the lattice points of a space lattice, a *crystal structure* is obtained. For any one space lattice there could be an unlimited number of crystal structures, each having different atoms or different arrangements of atoms associated with each lattice point.

A crystal structure is described, firstly by specifying the type of space lattice, for example in terms of the spacings and relative directions of the three sets of intersecting planes, and then by specifying the arrangement of the atoms associated with each lattice point.

Thus in the space lattice given by the intersections of the lines shown in Figure 7.1, each line being an intersection of two planes, we define axes OX, OY and OZ along three of the lines, the angles between the axes being α, β and γ;

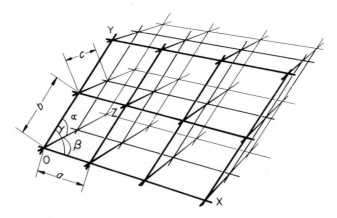

Figure 7.1. Lengths and angles specifying a space lattice

where α is the angle between the OY and OZ axes. We take the distances between the lattice points along each of the three axes, i.e. a, b and c, respectively, as the unit lengths in the three directions.

The space lattice is defined uniquely by the values of a, b, c and α, β, γ. The network of planes divides the region into prisms called *unit cells*. In Figure 7.1 the three unit lengths are shown as unequal and the three angles are also unequal and none equal to a right-angle. In some systems, a higher degree of symmetry exists and it is often more convenient to consider a larger unit cell. The smallest unit cell of a system is then referred to as a *primitive cell* and the larger unit is defined by appropriate values for a, b, c and α, β, γ.*

It has been shown that even allowing for all possible degrees of symmetry, there are only fourteen ways of arranging points in space to meet the fundamental description of a space lattice given earlier. The fourteen space lattices are shown in Figure 7.2, the degrees of symmetry present in each being given in Table 7.1 (see p. 102).

---

* Some authors use the terms *structure cell* and unit cell to denote the unit cell and primitive cell, respectively, as defined here.

The Crystalline State

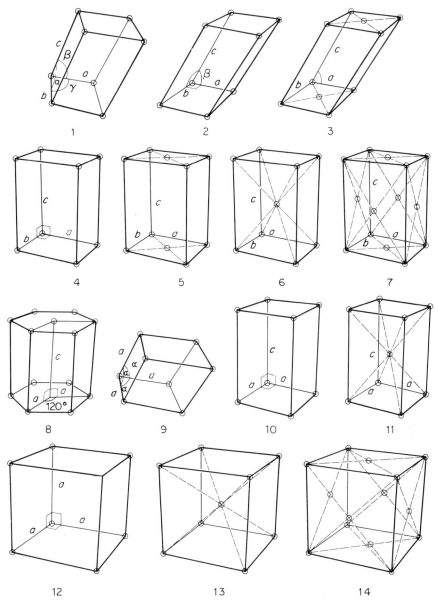

Figure 7.2. Unit cells of each of the 14 space lattices: (1) triclinic, simple; (2) monoclinic, simple; (3) monoclinic, base centred; (4) orthorhombic, simple; (5) orthorhombic, base-centred; (6) orthorhombic, body-centred; (7) orthorhombic, face-centred; (8) hexagonal; (9) rhombohedral; (10) tetragonal, simple; (11) tetragonal, body-centred; (12) cubic, simple; (13) cubic, body-centred; (14) cubic, face-centred

The arrangement of atoms in a crystal structure is usually specified by giving the coordinates of each atom in one unit cell with respect to the three axes that define the edges of the unit cell, the origin being at one corner of the unit cell. An example in two dimensions is given in Figure 7.3. Each unit cell in a crystal

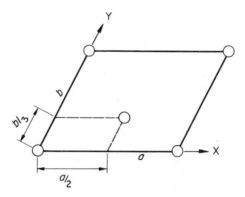

Figure 7.3. Example of atoms in a two-dimensional unit cell. Coordinates of atoms in the unit cell are $(0, 0)$ and $(\frac{1}{2}, \frac{1}{3})$. The atoms at $(1, 0)$, $(0, 1)$ and $(1, 1)$ are the $(0, 0)$ atoms of other unit cells

structure is identical in size, shape and orientation and in the positions of the atoms within it with every other unit cell in the same structure. Hence a complete description of one unit cell is a description of the structure.

By means of X-ray diffraction methods it is possible to determine the shape and size of the unit cell and also the distribution of atoms within it.

7.3 Indices of planes

In discussing crystal structures, it is often necessary to refer to various planes and directions relative to the crystal axes. The usual method is to specify them in terms of the *Miller index notation.*

A plane is defined by the length of its intercepts on the three crystal axes (the three edges of a unit cell), measured from the origin of coordinates. The intercepts are expressed in terms of the dimensions of the unit cell, which are the unit distances along the three axes. The reciprocals of these intercepts reduced to the smallest three integers having the same ratio are known as *Miller indices*.

As the origin may be taken at any lattice point, the lengths of the intercepts for a given plane are not specified, but their ratio is constant. Any other parallel plane would have the same indices, so that a particular set of Miller indices defines a set of parallel planes.

The Crystalline State

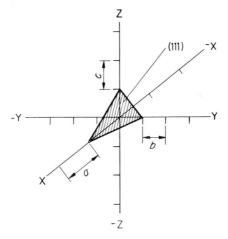

Figure 7.4. Intercepts of plane with crystal axes

The shaded plane in Figure 7.4 has intercepts 1, 1, 1 and therefore indices (111). A plane with intercepts 2, ∞, 1 (i.e. parallel to the OY axis) has reciprocal intercepts $\frac{1}{2}$, 0, 1 and Miller indices (102). If a plane cuts an axis on the negative side of the origin, the corresponding index will be negative and written with a line above the number. Some examples of indices of planes are shown in Figure 7.5.

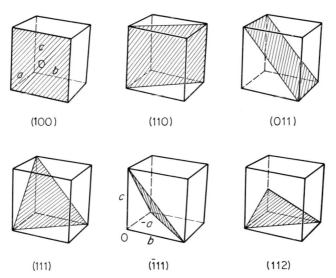

Figure 7.5. Examples of Miller indices of planes

Parentheses ( ) around a set of indices are used to signify a single set of parallel planes.

One frequently needs to specify collectively all the sets of planes which are exactly equivalent to one another. Thus in the cubic system which has a very high degree of symmetry, the plane (110) is equivalent to (101), (011), (1$\bar{1}$0), (10$\bar{1}$) and (01$\bar{1}$). To specify such sets of planes or planes of a *form*, the indices are enclosed in curly brackets or braces { }. Thus {110} will include the six indices stated above.

Reversal of the signs of all the indices merely denotes another parallel plane. Thus ($\bar{1}\bar{1}$0) is parallel to (110) and (10$\bar{1}$) to ($\bar{1}$01).

An alternative indexing system is frequently used for the hexagonal system. This is the *Miller–Bravais* system in which four axes with unit lengths a_1, a_2, a_3 and c as shown in Figure 7.6 are used. Then four numbers h, k, i and l are used

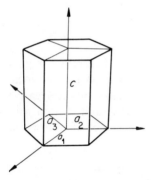

Figure 7.6. Relationship of Miller–Bravais indices to hexagonal cell

in each set of indices. Otherwise the treatment is the same as for Miller indices. The first three indices are not independent but must satisfy

$$h + k + i = 0$$

Conventional three-number Miller indices for a hexagonal lattice merely differ by omitting the third number. With four-number indices, the relationship of planes to the symmetry of the hexagonal lattice is more obvious and planes of a form have the same sets of indices, though in different orders. Thus the planes (10$\bar{1}$0), (01$\bar{1}$0) and (1$\bar{1}$00) are of the same form, but this would not be immediately obvious when using three-number indices: (100), (010) and (1$\bar{1}$0). Three-number indices have been used in the remainder of the book.

7.4 Indices of direction

A direction is defined in terms of the successive motions parallel to each of the three axes which are necessary to move from the origin to another point which

The Crystalline State

lies in the required direction. Suppose that the moves are a distance u times the unit distance a along the X axis, v times b parallel to the Y axis and w times c parallel to the Z axis, where u, v and w are the smallest set of integers that will perform the movement, then they are the indices of the direction and are written in square brackets thus: $[uvw]$. The X axis will be $[100]$, the negative Y axis $[0\bar{1}0]$, etc. Several directions are shown in Figure 7.7.

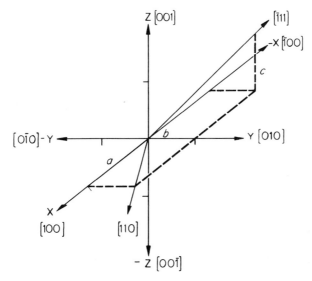

Figure 7.7. Examples of Miller indices of direction

A full set of equivalent directions (i.e. directions of a form) are indicated by carets $\langle \ \rangle$. Thus in the cubic system,

$$\langle 111 \rangle = [111] + [11\bar{1}] + [1\bar{1}1] + [\bar{1}11]$$

It is important to note that reciprocals are *not* used in computing directions. In a cubic system *only*, the $[hkl]$ direction is perpendicular to the (hkl) plane.

7.5 Crystal systems

As was shown in Figure 7.2, each space lattice is represented by various degrees of symmetry in the directions of the axes, the unit lengths along them and by extra lattice points within the unit cell. There are seven combinations of axis directions and unit cell dimensions known as the seven crystal systems which are given in Table 7.1.

Table 7.1 The seven crystal systems

System			Space lattice
Triclinic	$a \neq b \neq c$	$\alpha \neq \beta \neq \gamma \neq 90°$	Triclinic, simple
Monoclinic	$a \neq b \neq c$	$\alpha = \gamma = 90° \neq \beta$	Monoclinic, simple
			Monoclinic, end-centred
Orthorhombic	$a \neq b \neq c$	$\alpha = \beta = \gamma = 90°$	Orthorhombic, simple
			Orthorhombic, end-centred
			Orthorhombic, face-centred
			Orthorhombic, body-centred
Tetragonal	$a = b \neq c$	$\alpha = \beta = \gamma = 90°$	Tetragonal, simple
			Tetragonal, body-centred
Cubic	$a = b = c$	$\alpha = \beta = \gamma = 90°$	Cubic, simple
			Cubic, face-centred
			Cubic, body-centred
Hexagonal	$a = b \neq c$	$\alpha = \beta = 90°, \gamma = 120°$	Hexagonal, simple
Rhombohedral	$a = b = c$	$\alpha = \beta = \gamma \neq 90°$	Rhombohedral, simple

For each crystal system there are one or more types of space lattice distinguished by the differing degrees of symmetry of the unit cell. In all 'simple' lattices, the primitive cell is also the unit cell.

The crystal structures and the lattice constants of the elements are given in Table 1.1. It will be seen that each element adopts a crystal structure with a fairly high degree of symmetry, more than half adopting one of three structures, the face-centred and body-centred lattices of the cubic system and the hexagonal close-packed structure, which is a special example of the simple hexagonal lattice. The features of these three are described in the following sections.

7.6 Face-centred cubic structure

A unit cell of the face-centred cubic lattice is shown in Figure 7.8. Metals with this structure have an atom at each lattice point and nowhere else. By extending the structure as in Figure 7.9 it can be seen that atoms at the centres of faces of unit cells have identical surroundings to corner atoms and so can be taken as the corners of a different network of unit cells. Figure 7.10 shows a model of a cluster of spheres packed in a f.c.c. manner.

The coordinates of the atoms in one unit cell are $(0, 0, 0)$, $(\frac{1}{2}, 0, 0)$, $(0, \frac{1}{2}, 0)$ and $(0, 0, \frac{1}{2})$, i.e. there are four atoms per unit cell.

All the {111} planes are planes of closest packing as may be seen clearly in Figure 7.11. The spheres touch along the $\langle 110 \rangle$ directions, i.e. these are close-

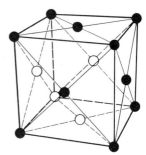

Figure 7.8. Face-centred cubic unit cell. The circles represent the nuclei of atoms

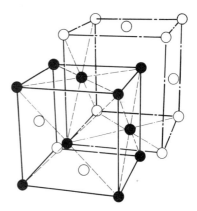

Figure 7.9. Two possible sets of unit cells for a face-centred cubic structure

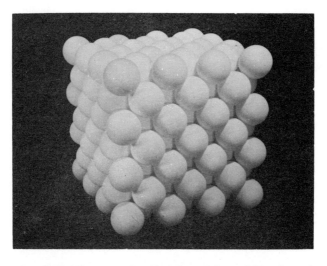

Figure 7.10. Face-centred cubic structure formed by spheres in contact

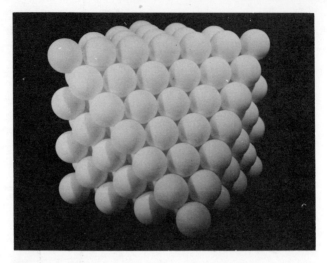

Figure 7.11. Face-centred cubic structure exposed on (111) plane showing close packing of atoms

packed directions. It will be seen that in a (111) plane each atom has six atoms which form a hexagon around it. There will also be three atoms in each of the adjacent parallel planes, which are at the same distance from it, so that each atom has twelve equidistant nearest neighbours (the closest possible packing of spheres).

7.7 Body-centred cubic structure

A unit cell of the body-centred cubic lattice is shown in Figure 7.12. There are two atoms per unit cell with coordinates $(0, 0, 0)$ and $(\frac{1}{2}, \frac{1}{2}, \frac{1}{2})$.

Figure 7.12. Body-centred cubic unit cell

Figures 7.13 and 7.14 show models of spheres packed in this manner. It can be seen from the second picture that each atom has eight equidistant nearest

The Crystalline State

Figure 7.13. Body-centred cubic structure formed by spheres in contact

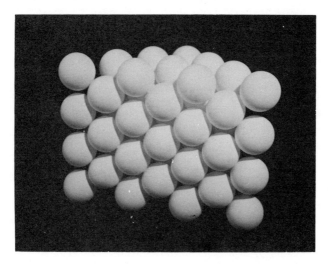

Figure 7.14. Body-centred cubic structure exposed on (110) plane showing close-packed [1$\bar{1}$1] and [$\bar{1}$11] directions

neighbours which touch along the [111] directions. Hence all the $\langle 111 \rangle$ directions are close-packed but there is no close-packed plane.

106   *Properties of Materials for Electrical Engineers*

7.8 Close-packed hexagonal structure

In the close-packed hexagonal structure close-packed layers are stacked on top of one another, but in a different sequence from that in a face-centred cubic lattice, giving the structure shown in Figure 7.15. The unit cell is shown in

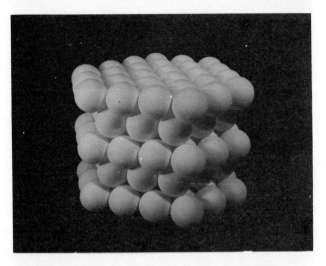

Figure 7.15. Hexagonal close-packed structure formed by spheres in contact

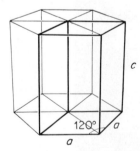

Figure 7.16. Relationship of unit cell to hexagon in hexagonal close-packed structure

relation to the hexagonal pattern in Figure 7.16. The atom sites in relation to the hexagonal prism are shown in Figure 7.17 from which it can be seen that there are two atoms per unit cell with coordinates $(0, 0, 0)$ and $(\frac{1}{3}, \frac{2}{3}, \frac{1}{2})$. This structure differs from the two described previously in that only one of the atoms per unit cell is at a lattice point.

The Crystalline State

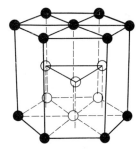

Figure 7.17. Atom sites of close-packed hexagonal system

7.9 The diamond structure

Diamond is one of the crystalline forms adopted by carbon. Each atom has four nearest neighbours so arranged that they lie at the corners of a regular tetrahedron of which the first atom lies at the centre, the bonding being tetrahedral covalent as described in Section 5.6. A ball-and-spoke model of the structure is shown in Figure 7.18. It can also be considered as two interpenetrating

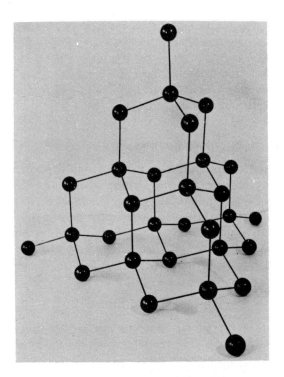

Figure 7.18. Structure of diamond

face-centred cubic lattices. The unit cell is chosen as a cube containing eight atoms as shown in Figure 7.19.

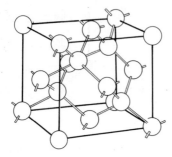

Figure 7.19. Cubic unit cell of diamond

7.10 Ionic crystals

The structure adopted by any particular ionic crystal depends upon the relative valencies and relative sizes of the various ions. The simplest to consider are those which consist of two elements with equal positive and negative valencies. These are typified by a formula of the type MX where M and X denote the metallic and non-metallic elements, respectively. Most of these compounds crystallize in one of four characteristic structures which are shown in Figure 7.20. Some, such as zinc sulphide, have different structures at different temperatures. Reference will be made to this compound in Section 12.11 when discussing pyroelectric materials.

7.11 Relation of crystal structure to properties

While some physical properties of materials are dependent strongly upon the nature of interatomic bonds and spacing but less so on the actual atomic arrangement, others depend strongly upon the exact crystal lattice adopted. This dependence can be properly understood only in terms of the wave-mechanical picture of atomic interaction. The rudiments of this have been given in Chapter 5 but further wave-mechanical concepts necessary for this understanding will be developed later.

Questions for Chapter 7

1. By considering a crystalline structure explain the difference between a *lattice point* and an *atom site* and between a *unit cell* and a *primitive cell*. [MST]

2. Calculate the nearest distance between the centres of two atoms and the density of nickel from the relevant information in Table 1.1.

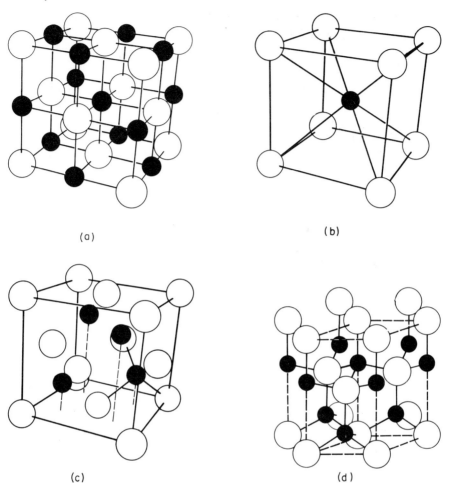

Figure 7.20. Structure of simple ionic compounds: (a) sodium chloride, (b) caesium chloride, (c) zincblende and (d) wurtzite. Each consists of two interpenetrating lattices: (a) two face-centred cubic, (b) two simple cubic, (c) two face-centred cubic with a different relative displacement than (a), and (d) two hexagonal close-packed

3. For cobalt:
   a. Calculate the density at 20 °C.
   b. Calculate the angle between the (111) and the (001) planes and the angle between the [111] direction and the (001) plane. [P]

4. Calculate the distances between successive (100), (110) and (111) planes in gold.

5. What type of bonding would you expect to find in the following substances: NaF, H_2, CH_4, MgO, $AlCl_3$?

Using these or other substances as examples, discuss the factors which control the type of structure found in ionic and in covalent solids. [P]

6. Certain physical properties of crystals of carborundum (SiC), naphthalene ($C_{10}H_8$) and cryolite (Na_3AlF_6) can be predicted on the basis of the atomic structure of the constituent elements and the types of interatomic bonding suggested by the formulae.

Enumerate these properties and explain the basis of the predictions.

[MST]

CHAPTER EIGHT

Solid Compounds

8.1 Introduction

Chemical compounds, i.e. materials comprised of at least two different elements, can be classed broadly into inorganic and organic compounds. The organic compounds are those that contain carbon and usually one or more of the elements hydrogen, oxygen and nitrogen, together sometimes with halogens and other non-metallic elements. All other compounds are known as inorganic and include metallic carbides and carbonates.

Many compounds of both classes are used by the electrical engineer—some because they have excellent insulating properties and others because they have some special electrical or magnetic property, for example semiconducting, piezoelectric or ferrimagnetic. Some general features of the two classes of compound are discussed in the remainder of this chapter, particular compounds with special properties being mentioned in the appropriate chapters later.

8.2 Ceramics

The term *ceramics* denotes those products which are made from inorganic crystalline materials and which have non-metallic properties. Simple examples are ionically bonded *magnesia*, MgO, and covalently bonded silicon carbide, SiC. The crystal structures of these are similar to those of sodium chloride (Figure 7.20(a)) and diamond (Figure 7.18), respectively.

An important group of ceramics is the silicates, consisting of metal ions and negative ions which contain silicon combined covalently with oxygen. Each silicon atom links to four oxygen atoms which are arranged in a regular tetrahedron with the silicon at the centre, as in Figure 8.1. Each oxygen atom requires one more electron to complete a stable shell of eight and can acquire it in one of two ways: either, it can receive an electron from a metal atom thereby forming an ion pair, or, it can form a covalent bond with another silicon atom. Two silicon atoms rarely have more than one oxygen atom in common, i.e. the tetrahedra touch only at the corners and not along edges or sides.

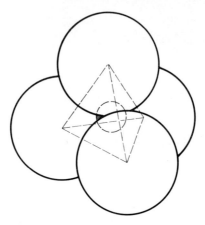

Figure 8.1. Basic unit of silicates. Silicon atom at centre of tetrahedron of four oxygen atoms

It is convenient to think of these silicon–oxygen groups in terms of the oxygen tetrahedra. Each corner of a tetrahedron either is joined to a corner of another tetrahedron or carries a negative charge. The structures commonly encountered are regular arrangements in chains, layers or a complete three-dimensional network. The types of chain and layer structures are illustrated in Figure 8.2.

Silica, SiO_2, has a network structure. It is most easily visualized as layers of the type shown in Figure 8.2(d) but in each layer the tetrahedra point alternately up and down. It exists in three allotropic forms, each stable over a different temperature range. A change from one form to another involves a sudden expansion or contraction, i.e. a dilatation.

Certain silicates will be mentioned in later chapters, e.g. kaolin, which is a layer structure of silicate combined with Al^{3+} ions and OH^- ions as shown diagrammatically in Figure 8.3.

Whereas ceramic materials are crystalline, some do not crystallize from the melt except at very slow rates of cooling, but instead increase steadily in viscosity as they cool and form *glasses*. This class of materials is discussed further in Section 8.4.

8.3 Physical properties of ceramics

Under applied loads, plastic deformation can occur in single crystals of a few of the simpler ceramics, but in polycrystalline specimens the ductility is suppressed so that, with very few exceptions like silver chloride, the materials are brittle.

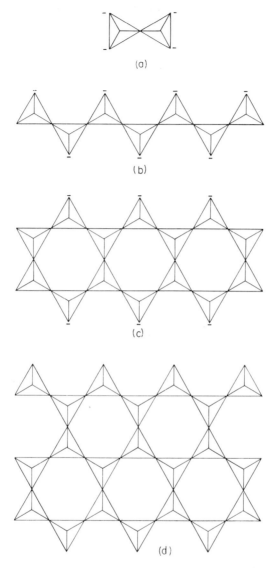

Figure 8.2. Tetrahedra networks found in silicates: (a) disilicate $(Si_2O_7^{6-})$, (b) single chain $(SiO_3^{2-})_n$, (c) double chain $(Si_4O_{11}^{6-})_n$, (d) layer $(Si_2O_5^{2-})_n$. Plan views are shown. All vertices facing out of the diagram as well as those marked — carry a negative charge

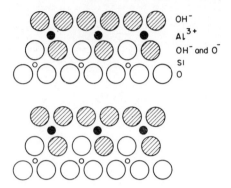

Figure 8.3. Structure of kaolin ($Al_2(OH)_4Si_2O_5$). The silicate layer combined with Al^{3+} and OH^- ions forms a sheet structure which is overall neutral but has a non-symmetrical charge distribution. The electrical polarization gives strong secondary bonding between layers

Most ceramic materials have a more complex crystal structure so that even single crystals of them tend to be brittle.

Ceramic materials have entirely primary bonding between atoms and so the melting points will be high. In compounds with di- and trivalent ions, the bonding is stronger than in compounds with only monovalent ions and the melting points will be correspondingly higher. Thus, sodium chloride melts at 801 °C, magnesia at 2800 °C and alumina, Al_2O_3, at 2000 °C. Such materials as the last two are excellent refractories for lining high-temperature furnaces, etc. Silica, with its allotropic changes and associated dilatation (see Section 8.2) would not be suitable for this purpose under conditions of rapidly varying temperature.

Ceramics, having no free electrons, are generally good thermal insulators and the thermal conductivity is generally decreased still more by crystal boundaries and pores in the structure. The conductivity may, however, increase considerably at high temperatures.

In ceramics which have low thermal conductivities, any localized change in temperature will not be followed rapidly elsewhere, so that thermal gradients are set up. These would lead to differential thermal expansion, which causes thermal stresses and may lead to cracking because of the brittle nature of these materials.

The two crystalline forms of carbon, diamond and graphite, which are usually considered with ceramics, are exceptional in having high thermal conductivities. Graphite is also unique in being highly anisotropic in this property. In graphite,

Solid Compounds

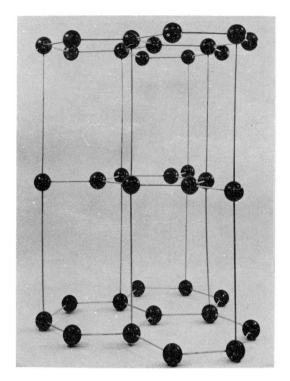

Figure 8.4. The structure of graphite. Interatomic distance within each layer = 1·42 Å. Distance between layers = 3·35 Å

the atoms are arranged in parallel hexagonal layers (Figure 8.4) having strong bonds within each layer and only weak Van der Waals forces between layers. The thermal conductivity in the plane of the layers is 100 times that in the perpendicular direction.

Graphite also has a highly anisotropic electrical conductivity. Apart from this material and a few semiconductors like silicon carbide (see Section 14.1.2), most ceramics have extremely high electrical resistivities below 300 °C. Their use as electrical insulators is discussed in Section 12.8.

Ceramics are chemically resistant to most acids, alkalis and organic solvents and are unaffected by oxygen, so that they are very durable.

8.4 Vitreous structures

Certain inorganic materials have very low rates of nucleation of crystals from the melt so that at normal rates of cooling they form glasses or vitreous structures. The commonest of these materials are silica, SiO_2, boric oxide, B_2O_3, and phosphorus pentoxide, P_2O_5. The atoms in a glass are linked in a three-dimensional manner as in a crystal but the structure is not regular, only short-term order existing. Figure 8.5 shows the manner in which the SiO_4 tetrahedra might be arranged in a silica glass.

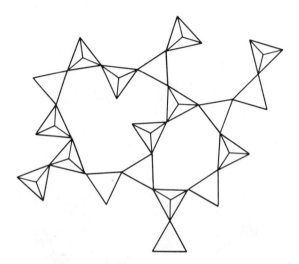

Figure 8.5. Representation of silica glass. A two-dimensional layer shows varying numbers of tetrahedra in each ring. Clear triangles represent tetrahedra with apices down, others with apices up. The true structure will also be random in the third dimension

Pure silica glass has a low coefficient of thermal expansion and a high softening temperature. Also the dilatation found with the crystalline forms (see Section 8.2) does not occur in the vitreous state. All these factors give it good thermal shock resistance.

The oxides of the alkali metals are incapable of forming glasses themselves, but will combine with vitreous silica. Some of the bridging oxygen bonds between silicon atoms break and each is replaced by two non-bridging oxygen atoms each carrying a negative charge. The metal ions preserve electrical neutrality as a whole.

$$-O-\underset{\underset{O}{|}}{\overset{\overset{O}{|}}{Si}}-O-\underset{\underset{O}{|}}{\overset{\overset{O}{|}}{Si}}-O- + Na_2O \rightarrow -O-\underset{\underset{O}{|}}{\overset{\overset{O}{|}}{Si}}-O^- \;\; O^- -\underset{\underset{O}{|}}{\overset{\overset{O}{|}}{Si}}-O- + 2\,Na^+$$

Solid Compounds 117

The breaks in the structure cause a weakening so that the softening temperature of the glass is lowered, enabling it to be moulded to shape at lower temperatures than for pure silica glass. The thermal expansion is, however, increased by the structural modifications so that the thermal shock resistance is decreased. Further reference to glazed articles will be found in Section 12.8.

8.5 Glass-ceramics

Glasses are formed by cooling liquids which have a very low crystal nucleation rate. By introducing suitable catalysts, materials formed in the glassy state can be devitrified by subsequent heat treatment, i.e. they become partially or entirely crystalline. The products are known as *glass-ceramics*.

The mechanical strengths of glass-ceramic articles are considerably greater than those of similar articles made of the same material either in the vitreous state or as a ceramic by pressing and sintering powder. Also many physical properties are superior and can be varied at will over wide ranges by slight changes of composition. As an example, materials with coefficients of thermal expansion ranging from small negative to high positive values are possible.

Glass-ceramic articles can be shaped by conventional glass manufacturing processes and do not change dimensions or surface finish appreciably during devitrification.

8.6 Organic compounds

The covalent bonding of carbon atoms with four bonds arranged in a tetrahedral manner has been discussed in Section 5.6. Carbon atoms link to one another in long chains and in ring structures so that there is a limitless number of possible organic compounds. Organic solids consist of molecules which are held to each other by weak secondary or Van der Waals bonds. The larger the molecule in a compound, the greater is the total force between molecules, so that strength is generally greater and the melting or softening point is higher.

Most naturally-occurring organic materials and the large number of synthetic ones which are of engineering importance consist of large molecules. The properties can be properly understood only in terms of the molecular structures, which will be discussed in the following sections.

8.7 Saturated hydrocarbons

These are compounds in which the carbon atoms in each molecule are joined by single covalent bonds and any remaining carbon bonds are completed by hydrogen atoms.

The carbon atoms may all be in a single chain or there may be side branching. The two simplest members of the series, methane (CH_4) and ethane (C_2H_6), have already been described in Section 5.4. Though they are three-dimensional structures, it is convenient to represent them by two-dimensional structural formulae, which for these two and the next member, propane (C_3H_8), are:

$$\begin{array}{ccc} \text{H} & \text{H H} & \text{H H H} \\ | & |\ | & |\ |\ | \\ \text{H}-\text{C}-\text{H} & \text{H}-\text{C}-\text{C}-\text{H} & \text{H}-\text{C}-\text{C}-\text{C}-\text{H} \\ | & |\ | & |\ |\ | \\ \text{H} & \text{H H} & \text{H H H} \\ \text{methane} & \text{ethane} & \text{propane} \end{array}$$

Each member is derived from the previous one by replacing a hydrogen atom with a *methyl* ($-CH_3$) group. By repeating this process each time at the end of the carbon chain, we get a *homologous series* with the general formula C_nH_{2n+2}. This may be written in a more informative manner as $CH_3 \cdot (CH_2)_{n-2} \cdot CH_3$ or in a displayed form as

$$H-\underset{H}{\overset{H}{\overset{|}{C}}}\left(\underset{H}{\overset{H}{\overset{|}{C}}}\right)_{n-2}\underset{H}{\overset{H}{\overset{|}{C}}}-H$$

The substitution of a methyl group for a hydrogen atom need not be carried out at the end of the chain, so that branched compounds can occur. Thus butane (C_4H_{10}) has two forms or *isomers*

$$\begin{array}{cc} \text{H H H H} & \text{H H H} \\ |\ |\ |\ | & |\ |\ | \\ \text{H}-\text{C}-\text{C}-\text{C}-\text{C}-\text{H} & \text{H}-\text{C}\text{------}\text{C}\text{------}\text{C}-\text{H} \\ |\ |\ |\ | & |\ |\ | \\ \text{H H H H} & \text{H}\ |\ \text{H} \\ & \text{H}-\text{C}-\text{H} \\ & | \\ & \text{H} \\ \text{normal butane} & \text{iso-butane} \end{array}$$

Each higher member of the series will have more isomers.

This series is known as the paraffins. The larger the molecule, i.e. larger n, the greater is the bonding between molecules so that the melting point and boiling point are higher and, for solids, the strength is greater. The intermolecular forces are such that the lowest members of the series, those up to pentane (C_5H_{12}), are gases at room temperature and atmospheric pressure, the medium ones are liquids (oils) while the highest ones are solids (paraffin waxes).

Solid Compounds

The paraffins do not react easily, the only notable reactions being:

a: combination with oxygen to give carbon dioxide and water (a strongly exothermic reaction often used as a primary source of energy)
b: substitution of chlorine for hydrogen in a chlorine atmosphere and diffused light. This can happen to various degrees, for example methane reacts with chlorine to give a mixture of methyl chloride CH_3Cl, methylene chloride CH_2Cl_2, chloroform $CHCl_3$ and carbon tetrachloride CCl_4.

A wide range of compounds is formed by replacing hydrogen with other atoms or radicals, not necessarily directly, by a series of chemical reactions that do not concern us here. Some of the most common groups or radicals in organic compounds are

$$-OH \qquad\qquad \underset{\underset{O}{\|}}{-C-OH} \qquad\qquad -NH_2$$
$$\text{alcohol} \qquad\qquad \text{acid} \qquad\qquad \text{amine}$$

$$\underset{\underset{O}{\|}}{-C-H} \qquad\qquad \underset{/}{\overset{\backslash}{}}C=O \qquad\qquad -O-$$
$$\text{aldehyde} \qquad\qquad \text{ketone} \qquad\qquad \text{ether}$$

8.8 Unsaturated hydrocarbons

The formation of a double bond between two carbon atoms was discussed in Section 5.6, the simplest hydrocarbon exhibiting this being ethylene, C_2H_4, which can be represented in displayed form as

$$\begin{array}{c} H \quad H \\ | \quad\; | \\ C=C \\ | \quad\; | \\ H \quad H \end{array}$$

As with the paraffin series, homologues are formed by substitution of methyl groups for hydrogen atoms. The next members of the *olefine* series, which have the general formulae C_nH_{2n} are *propylene* or *propene*

$$\begin{array}{c} \quad\; H \quad H \quad H \\ \quad\; | \quad\; | \quad\; | \\ H-C-C=C \\ \quad\; | \quad\quad\; | \\ \quad\; H \quad\quad H \end{array}$$

and the three isomers of *butylene*

$$\underset{\text{1-butene}}{\begin{array}{c} \text{H H H H} \\ \text{H–C–C–C=C} \\ \text{H H H} \end{array}} \quad \underset{\text{2-butene}}{\begin{array}{c} \text{H H H H} \\ \text{H–C–C=C–C–H} \\ \text{H H} \end{array}} \quad \underset{\text{iso-butene}}{\begin{array}{c} \text{H CH}_3 \\ \text{C=C} \\ \text{H CH}_3 \end{array}}$$

The double bonds are relatively weak and revert easily to single bonds by taking part in addition reactions such as the reaction of ethylene with chlorine to give ethylene chloride

$$\begin{array}{c} \text{CH}_2 \\ \| \\ \text{CH}_2 \end{array} + \text{Cl}_2 \rightarrow \begin{array}{c} \text{CH}_2\text{Cl} \\ | \\ \text{CH}_2\text{Cl} \end{array}$$

8.9 Aromatic hydrocarbons

The paraffin hydrocarbons and the wide range of compounds that can be derived from them by substitution of various organic radicals for hydrogen atoms are known as *aliphatic* compounds.

By contrast, the *aromatic* compounds are those containing six-ring carbon structures and which have strong distinctive smells. The simplest is benzene (C_6H_6) in which the carbon atoms are arranged in a regular hexagon with the hydrogen atoms outside

$$\begin{array}{c} \text{H} \\ \text{C} \\ \text{HC} \quad \text{CH} \\ | \quad\quad | \\ \text{HC} \quad \text{CH} \\ \text{C} \\ \text{H} \end{array}$$

To satisfy the quadrivalence of carbon, the benzene ring is considered to have one primary or σ bond between each pair of adjacent carbon atoms and *p* orbitals (see Section 5.6), one on each carbon atom, forming π bonds as in ethylene. The *p* orbitals overlap with the carbon atoms on both sides so that the structure can be said to be a resonance between

⬡ and ⬡

In structural formulae, the benzene ring is usually drawn as a simple hexagon, the carbon and hydrogen atoms not being shown. Any group substituted for a hydrogen atom is indicated adjacent to the appropriate corner of the ring. Some

Solid Compounds

examples are:

$$\underset{\text{toluene}}{\text{C}_6\text{H}_5\text{-CH}_3} \quad \underset{\text{phenol}}{\text{C}_6\text{H}_5\text{-OH}} \quad \underset{\textit{ortho}\text{-xylol}}{\text{C}_6\text{H}_4(\text{CH}_3)_2} \quad \underset{\textit{meta}\text{-xylol}}{\text{C}_6\text{H}_4(\text{CH}_3)_2} \quad \underset{\textit{para}\text{-xylol}}{\text{C}_6\text{H}_4(\text{CH}_3)_2}$$

8.10 Polymerization

The formation of very large molecules by repeated combination of simple molecules is known as *polymerization*. One of the simplest reactions of this type is found with ethylene under suitably high pressure and temperature and in the presence of a catalyst. The weaker part of each double bond breaks and the separate molecules combine to form long-chain paraffin molecules.

$$\begin{array}{c} \text{H H} \\ | \; | \\ \text{C}=\text{C} \\ | \; | \\ \text{H H} \end{array} + \begin{array}{c} \text{H H} \\ | \; | \\ \text{C}=\text{C} \\ | \; | \\ \text{H H} \end{array} \rightarrow \begin{array}{c} \text{H H H H} \\ | \; | \; | \; | \\ -\text{C}-\text{C}-\text{C}-\text{C}- \\ | \; | \; | \; | \\ \text{H H H H} \end{array}$$

The units of the *monomer*, in this case ethylene, have joined to form a *polymer*; this one is known as *polyethylene* or *polythene*.

Molecules of the monomer must be available in the region of the end of a chain if continued polymerization is to occur. Hence the reaction can continue easily until most of the monomer has been used up, after which unpolymerized molecules must diffuse to the regions where further combination can take place. Terminal radicals are necessary to provide stability; e.g. hydrogen peroxide is used to provide terminal —OH groups. The average size of molecule can to some extent be regulated by controlling the concentration of the compound providing the terminal radicals.

The *degree of polymerization* (D.P.) is the number of *mers* or repeating monomer units in the molecule, i.e. it is the ratio of the molecular weight of the polymer to the molecular weight of the monomer. A polymer will contain a mixture of molecular sizes, so that the D.P. represents an average value.

The structure of polythene is similar to that of a high molecular weight paraffin and has similar chemical properties, in particular being fairly inert. The long-chain molecules are linked by Van der Waals forces and so the melting points will be low relative to ceramics and metals. It does not have a sharply defined melting point, but softens in the range 100–150 °C (see Section 8.12).

Polymerization reactions are conveniently divided into two types: *addition polymerization*, such as that just described, where there are no by-products and *condensation polymerization* in which functional groups of the molecules react to form the polymer and also to eliminate a small molecule such as water.

The acid–base reaction of organic chemistry is an example of a condensation reaction, an alcohol (R_1OH) and an acid (R_2COOH) condensing to form an *ester* and water

$$R_1OH + R_2\underset{O}{\overset{\|}{C}}{-}OH \rightarrow R_1{-}O{-}\underset{O}{\overset{\|}{C}}{-}R_2 + H_2O$$

Here the R's are used to denote hydrocarbon radicals (e.g. methyl,CH_3- or ethyl,C_2H_5-), the reaction being unaffected by the exact composition and structures of these radicals.

8.11 Some common polymers

The number of polymers in commercial use is already very large. Some of the more important and the compounds from which they are formed are listed below.

Addition polymers

vinyl chloride → polyvinyl chloride (p.v.c.)

propylene → polypropylene

styrene → polystyrene

Solid Compounds

$$\underset{\text{methyl methacrylate}}{\overset{H}{\underset{H}{C}}=\overset{CH_3}{\underset{COOCH_3}{C}}} \rightarrow \underset{\text{polymethylmethacrylate (perspex)}}{-\overset{H}{\underset{H}{C}}-\overset{CH_3}{\underset{COOCH_3}{C}}-\overset{H}{\underset{H}{C}}-\overset{CH_3}{\underset{COOCH_3}{C}}-}$$

$$\underset{\text{tetrafluorothylene}}{\overset{F}{\underset{F}{C}}=\overset{F}{\underset{F}{C}}} \rightarrow \underset{\text{polytetrafluoroethylene}}{-\overset{F}{\underset{F}{C}}-\overset{F}{\underset{F}{C}}-\overset{F}{\underset{F}{C}}-\overset{F}{\underset{F}{C}}-}$$

Two or more types of monomer can be combined to form a copolymer. Vinyl chloride and vinyl acetate, for example, give

$$-\overset{H}{\underset{H}{C}}-\overset{H}{\underset{Cl}{C}}-\overset{H}{\underset{H}{C}}-\overset{H}{\underset{O.CO.CH_3}{C}}\text{———}\overset{H}{\underset{H}{C}}-\overset{H}{\underset{Cl}{C}}-$$

Condensation polymers

$$R_1\diagup^{CH_2OH}_{CH_2OH} + R_2\diagup^{COOH}_{COOH}$$

double alcohol   double acid

$$\rightarrow HO{\left[CH_2.R_1.CH_2-O-\underset{\overset{\|}{O}}{C}-R_2-\underset{\overset{\|}{O}}{C}-O\right]}_n H + H_2O$$

polyester
(e.g. terylene)

$$NH_2.R.COOH \rightarrow H[NH.R.CO]_n OH + H_2O$$

amino acid   polyamide
(e.g. nylon)

In addition to chain polymers, network polymers are formed where the monomer molecules each have three or more functional groups.

$$\text{phenol} + \text{formaldehyde} + \text{phenol} \rightarrow \text{phenol formaldehyde} + H_2O$$

Any of the H atoms on a phenol molecule, except that in the —OH group, can react in this manner so that one phenol molecule can be linked to one, two or three others. Spatial reasons would probably prevent four or five linkages to one molecule taking place. The polymer becomes a three-dimensional network structure.

$$\text{urea} + \text{formaldehyde} + \text{urea} \rightarrow \text{urea formaldehyde} + H_2O$$

Each urea molecule could make four linkages, spatial reasons permitting, so that urea formaldehyde is another network polymer.

8.12 Relationship of mechanical properties to polymer shape

The products of addition polymerization discussed above and some products of condensation polymerization are linear or chain molecules which will be held together only by Van der Waals forces. The exact nature of these forces depends upon the side groups present. In chains which are symmetric, like polythene and p.t.f.e., the molecules are non-polar, while unsymmetrical chains, like polyvinyl chloride, are strongly polar (see Section 5.8). Hence the force of attraction between chains of the latter will be considerably greater than that between chains of the former. Also, with more complicated shapes of side groups, it is less easy for molecular chains, which are intertwined, to slide past one another, thus increasing the resistance to plastic deformation.

While the mechanical behaviour of the various chain polymers differ in detail, they all have a common pattern. At very low temperatures they are hard, rigid, glassy materials giving limited elastic deformation and fracturing in a brittle manner. As the temperature is raised, they will exhibit viscoelastic strain, i.e. under a given load they gradually deform to a limited extent and on removal of the load they recover their original shape after a time. The tempera-

Solid Compounds

ture below which this viscoelastic behaviour is insignificant is the *glass-transition temperature*. At still higher temperatures, the materials will also undergo irreversible deformation under applied load, i.e. they flow. This mechanism becomes significant at the *flow temperature*.

Polyethylene is well above its glass-transition temperature at room temperature and behaves in a rubbery manner on stressing with a delayed recovery on unloading. The flow temperature is a little above 100 °C.

Polymers with large side groups such as polystyrene and the acrylic plastics (polymethyl acrylate and polymethyl methacrylate) are below their glass-transition temperatures at room temperature and so are hard and brittle. P.t.f.e. is a white translucent plastic of stiff rubbery consistency at room temperature.

These materials, which are viscoelastic at intermediate temperatures and fluids at higher temperatures are called *thermoplastic resins*. Articles to be made from these resins can be shaped by mechanically deforming material above the flow temperature and cooling before removal of the restraining forces. Thermoplastics can be joined by welding, i.e. heating the contact area and then mechanically pressing the parts together while they cool.

Polymers such as phenol formaldehyde and urea formaldehyde can have more than two linkages per molecule and so build up to three-dimensional networks which are continuous structures of covalent bonds. When completely polymerized, such materials will not undergo any irrecoverable plastic deformation, even at high temperatures and will fracture only in a brittle manner. They can be used at higher temperatures than most thermoplastic resins, the limit being the temperature at which decomposition or charring occurs. Articles of these resins are usually manufactured from partially polymerized material which is pressed into moulds and heated to complete polymerization. They are known as *thermosetting resins*.

With so many variations possible in the shape and size of polymer molecules, a wide range of mechanical properties are obtained.

8.13 Elastomers

Natural rubber is a polymer of a double-bonded compound called *cis*-isoprene

$$\begin{array}{c} \text{H} \quad \text{CH}_3 \quad \text{H} \quad \text{H} \\ | \quad\quad | \quad\quad | \quad | \\ \text{C}=\text{C}-\text{C}=\text{C} \\ | \quad\quad\quad\quad | \\ \text{H} \quad\quad\quad\quad \text{H} \end{array} \rightarrow \left[\begin{array}{c} \text{H} \quad \text{CH}_3 \quad \text{H} \quad \text{H} \\ | \quad\quad | \quad\quad | \quad | \\ \text{C}-\text{C}=\text{C}-\text{C} \\ | \quad\quad\quad\quad | \\ \text{H} \quad\quad\quad\quad \text{H} \end{array}\right]_n$$

The polymer itself has one double bond per mer and the methyl group and hydrogen atom attached to the unsaturated carbon atoms both lie on the same side of the chain. The double bond does not permit rotation but each single —C—C— bond can allow rotation so that a chain molecule can adopt many different configurations. Although the structure of polythene chains permits rotation at every —C—C— bond, the rubber and other polymers of similar type have a greater flexibility than polythene between the glass-transition temperature and the flow temperature.

Polymers of this type which have a large elastic range are known as *elastomers*. The various synthetic rubbers have a similar double-bond structure, as, for example, in neoprene which is a polymer of chloroprene.

$$\begin{array}{cccc} H & Cl & H & H \\ | & | & | & | \\ C=C-C=C \\ | & & | \\ H & & H \end{array} \rightarrow \left[\begin{array}{cccc} H & Cl & H & H \\ | & | & | & | \\ C-C=C-C \\ | & & & | \\ H & & & H \end{array} \right]_n$$

Plain chains of polyisoprene will slide fairly easily past one another when the material is deformed, but this flow can be suppressed while large elastic deformations can still occur by a small amount of cross-linking which will anchor the chains locally. The usual method of cross-linking is by *vulcanizing* with sulphur. A few of the unsaturated double bonds are broken and sulphur atoms form cross-links between chains

$$\begin{array}{c} \text{CH}_3\ \ \text{H} \\ |\ \ \ \ | \\ -\text{CH}_2-\text{C}=\!=\!\text{C}-\text{CH}_2- \\ \\ -\text{CH}_2-\text{C}=\!=\!\text{C}-\text{CH}_3- \\ |\ \ \ \ | \\ \text{CH}_3\ \ \text{H} \end{array} + 2\text{S} \rightarrow \begin{array}{c} \text{CH}_3\ \ \text{H} \\ |\ \ \ \ \ \ \ | \\ -\text{CH}_2-\text{C}\!-\!-\!-\!\text{C}-\text{CH}_2- \\ |\ \ \ \ \ \ \ | \\ \text{S}\ \ \ \ \text{S} \\ |\ \ \ \ \ \ \ | \\ -\text{CH}_2-\text{C}\!-\!-\!-\!\text{C}-\text{CH}_2- \\ |\ \ \ \ \ \ \ | \\ \text{CH}_3\ \ \text{H} \end{array}$$

With increased sulphur content, more cross-linking occurs, restricting the range of elastic deformation. When all the unsaturated bonds have been used up by complete vulcanization, giving ebonite, the rigidity reaches a maximum.

When exposed to light in the presence of oxygen, a similar cross-linking process occurs which makes the rubber hard and brittle. In this case, the bridging atoms are oxygen. Natural rubber is also attacked by oils, but neoprene is immune to this.

8.14 Epoxy resins

A special group of polymers formed by addition polymerization are the epoxy resins (Araldite, etc.). The starting materials contain an epoxy group

$$-\text{C}\underset{}{\overset{\text{O}}{\diagup\!\!\diagdown}}\text{C}-$$

Solid Compounds

which will react with an alcohol to give an ether and a further hydroxyl group. This hydroxyl group can in turn react with yet another epoxy group in the same way.

$$R \cdot OH + \underset{\diagdown O \diagup}{CH_2 - CH-} \rightarrow R \cdot O - CH_2 - \underset{OH}{\overset{|}{CH}} -$$

$$\underset{\diagdown O \diagup}{CH_2 - CH-} + R \cdot O - CH_2 - \underset{OH}{\overset{|}{CH}} - \rightarrow R \cdot O - CH_2 - \underset{O - CH_2 - \underset{OH}{\overset{|}{CH}} -}{\overset{|}{CH}} -$$

Various other radicals will also react to give polymerization.

These resins are thermosetting, but some will cure at room temperature once the monomer and setting agent are mixed. There is very little shrinkage as the mixture passes from the liquid to the solid state and there are high internal stresses. Also these differ from most thermosetting polymers in that there are no by-products.

The resulting materials have high adhesive strength due to the polarity of the hydroxyl and ether groups. They will also form chemical bonds with surfaces, e.g. metals, where active hydrogen may be found.

They are very inert chemically, being unaffected by caustic materials and are extremely resistant to acids.

The cross-linking points are more widely spaced than in phenol formaldehyde and similar resins and so give greater flexibility and toughness.

8.15 Fillers

The properties of pure polymers are frequently modified in some manner by the deliberate addition of filler compounds which are usually non-polymeric and insoluble. For example a reinforcing filler may be used to improve the strength. Up to 50 per cent. by weight of cellulose fibre materials, in the form of cloth, rags, wood or paper, is commonly used in thermosetting resins, the resulting strength depending on the form and proportion of the filler. Wood flour is frequently used in phenol formaldehyde (bakelite) mouldings to reduce both cost and density while having little effect on strength. Paper and linen are used in laminated structures, which are manufactured as sheets, tubes and blocks, to give greater strength and toughness.

Asbestos and mica can be used to increase the heat resistance and also have good electrical properties. Powdered minerals may be used to increase hardness, but a fibrous form must be used to increase tensile strength.

8.16 Plasticizers

These are compounds added either to aid the moulding process or to modify the final properties. They cause partial separation of the long-chain molecules so that the forces between them are decreased. This gives thermoplastic materials a more flexible or rubber-like nature, i.e. the glass-transition temperature is lowered. For example pure polyvinyl chloride is a hard substance at room temperature, but in the plasticized state is sufficiently flexible to be used as a covering for flexible cables.

Questions for Chapter 8

1. Arrange the following in the probable order of increasing melting temperatures, giving your reasons: KCl, CaO, B_2O_3.

2. Discuss in general terms the differences in structure and properties of ceramics, glasses and glass-ceramics.

3. 0·3 per cent. by weight of hydrogen peroxide is added to propylene before polymerization. If all the peroxide is used as terminal radicals for the polypropylene molecules, what is the degree of polymerization?

4. What is the molecular weight of polymethylmethacrylate which has a D.P. of 1500?

5. A rubber contains 95 per cent. chloroprene and 5 per cent. sulphur by weight. If all the sulphur is used for cross-linking during vulcanization, what fraction of the possible cross-links are joined?

   When exposed to ozone, further cross-linking by oxygen occurs. What will be the percentage increase in weight when the number of oxygen cross-links equals the number of sulphur cross-links?

6. Discuss the roles of fillers and plasticizers in commercial polymers.

CHAPTER NINE

Thermodynamics of Materials

9.1 Introduction

Mention has already been made in Chapter 6 of the effects of temperature upon certain properties of gases. The temperature variation of physical properties of matter in all states of aggregation will figure prominently in subsequent chapters. This variation is largely dependent upon the distribution of thermal energy among the electrons, atoms or molecules of a substance, a problem in statistical thermodynamics. In this chapter, the distribution and the manner in which it affects the rates of reactions are discussed.

9.2 Kinetic energy of a gas molecule

The kinetic energy of translation of the molecules of a perfect gas is equal to $\frac{1}{2}mn\overline{C^2}$ (see Section 6.3). Also

$$pV = \bar{R}T = \tfrac{2}{3} \times \tfrac{1}{2}mn\overline{C^2}$$

for one mole, so that the kinetic energy is $\frac{3}{2}\bar{R}T$ per mole or on average $\frac{3}{2}kT$ per molecule.

Now a gas molecule has three degrees of freedom for translational motion (e.g. the components of its velocity in three mutually perpendicular directions define its velocity). Maxwell has shown that in a system which has several degrees of freedom and obeys the laws of classical mechanics, the total energy of a system is equally divided among the different degrees of freedom. This is termed the principle of the *equipartition of energy*. Hence the kinetic energy of translation of a molecule of a gas in one degree of freedom is on average $\frac{1}{2}kT$.

Any rotational or internal vibrational motions that a molecule may have are also degrees of freedom which will share in the equipartition of energy.

For a diatomic molecule, the rotational and vibrational motions are quantized (see Section 3.6 for a treatment of a vibrational motion) so that the molecule can be excited in either of these modes only if it can receive the necessary discrete value of energy. Hence the rotational or vibrational modes will not generally be excited unless the average kinetic energy per degree of freedom, $\frac{1}{2}kT$, is at least of the same order of magnitude as the energy of the lowest excited state.

For example consider an oxygen molecule which can be approximated, as far as mass is concerned, to two point masses representing the nuclei (each of mass $16 \times 1.66 \times 10^{-27}$ kg) at a distance 2×10^{-10} m apart. Its greatest moment of inertia will be about an axis perpendicular to the line of centres (Figure 9.1) and will be

$$2 \times 16 \times 1.66 \times 10^{-27} \times (10^{-10})^2 \text{ kg m}^2 = 53 \times 10^{-47} \text{ kg m}^2$$

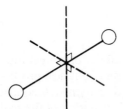

Figure 9.1. Diatomic molecule, showing principal axes of rotation

The least energy to excite a rotational motion about this axis would be, from equation (5.1),

$$E = \frac{1 \times 2 \times (6.625 \times 10^{-34})^2}{8\pi^2 \times 53 \times 10^{-47}} \text{ J}$$

$$= 2.1 \times 10^{-23} \text{ J}$$

As the average kinetic energy of a translational degree of freedom, $\frac{1}{2}kT$, is $0.7 \times 10^{-23} T$ J, the minimum rotational energy would be less than $\frac{1}{2}kT$ even for temperatures as low as 4 K. Hence the rotational degrees of freedom about axes perpendicular to the molecule axis will be excited, and at room temperature, the possible energy values will be so close that they act more like a classical continuous energy band.

However, since most of the mass of the molecule is concentrated in the nuclei, the moment of inertia about the axis through the line of centres will be of a much smaller order of magnitude, probably by a factor of 10^{-10}. The smallest rotational energy possible would be approximately 10^{-13} J, which is much greater than $\frac{1}{2}kT$ at any practical temperature. Hence this mode is unlikely to be excited due to ordinary intermolecular collisions. Also for such diatomic molecules, the energy of the lowest vibrational motion along the line of centres is too high for that motion to be excited at room temperature.

Therefore a diatomic molecule possesses five degrees of freedom that will be excited at all temperatures from its boiling point to very high values. By the

Thermodynamics of Materials

principle of equipartition of energy there will be, on average, $\frac{1}{2}kT$ per molecule per degree of freedom, so that the total energy is $\frac{5}{2}\overline{R}T$ per mole.

In a monatomic gas, the rotational modes will not be excited because the moment of inertia about any axis is so small. Hence the energy of translation is the total energy.

With more complicated molecules, the third rotational mode and internal vibrational modes may also be excited.

9.3 Specific heat of a gas

The specific heat capacity of a gas at constant volume (C_v) is the quantity of heat required to raise the temperature of unit quantity of the gas by one unit, while the volume remains unchanged. This heat is used entirely to increase the total kinetic energy of the molecules so that for one mole of a monatomic gas it should be $\frac{3}{2}\overline{R}$ and for one mole of a diatomic gas it should be $\frac{5}{2}\overline{R}$. A few examples to illustrate this are given in Table 9.1.

Table 9.1 Experimental and theoretical values of specific heat capacities for some gases (in J kmol^{-1} K^{-1})

Gas	Experimental value of C_v at room temperature		Theoretical value of C_v
Argon	12475	$\frac{3}{2}\overline{R} =$	12470
Helium	12475		
Air	20720		
Carbon monoxide	20680	$\frac{5}{2}\overline{R} =$	20780
Nitrogen	20635		

9.4 Specific heat of a solid

We have shown that the specific heat capacity of a gas depends upon the number of degrees of freedom of the molecules, which is three for a monatomic gas and five or more for all others.

The atoms in a solid are not mobile (except in so far as they migrate through the bulk by solid-state diffusion) but each vibrates about its mean position. The vibrational motion of an atom can be resolved into three mutually perpendicular components and each component requires two quantities to define it, e.g. the instantaneous values of kinetic and potential energies or the frequency and amplitude. Hence six independent terms are necessary to specify completely the motion, so that the atom has six degrees of freedom.

By the principle of equipartition of energy, the energy associated with each degree of freedom should be $\frac{1}{2}kT$ per atom on average or $\frac{1}{2}\bar{R}T$ per mole of atoms. The total energy should therefore be $6 \times \frac{1}{2}\bar{R}T = 3\bar{R}T$ so that $C_v = 3\bar{R}$ per mole.

On the basis of experimental results, Dulong and Petit had in 1819 formulated their law that the *atomic heat*, i.e. the product of specific heat capacity at constant volume per unit mass and atomic weight, is the same for all substances and is equal to 24 946 J kmol^{-1} K^{-1}.

While Dulong and Petit's law is approximately true for metals around room temperature, it does not apply to all materials at room temperature or to any material at low temperatures. All substances are found to have the same shape of curve for the variation of specific heat with temperature, which is shown in Figure 9.2. Near 0 K, the atomic heat is proportional to the cube of the absolute temperature and at high temperatures approaches $3\bar{R}$.

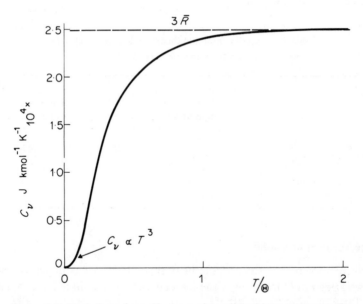

Figure 9.2. Variation of specific heat of a solid with temperature. By expressing temperature in the form T/Θ, where Θ is a characteristic temperature for each solid material, one curve applies to all materials

The fall-off at low temperatures is due to the failure to excite the vibrations when $\frac{1}{2}kT$ becomes small relative to the minimum vibrational energy, the vibrations being quantized.

Thermodynamics of Materials

A theoretical analysis by Debye based on quantum-mechanical principles applied to the possible vibrational modes in a crystal led to the formula*

$$C_v = 3\bar{R}\left[12\frac{T^3}{\Theta^3}\int_0^{\Theta/T}\frac{x^3\,dx}{e^x-1} - \frac{3\Theta/T}{e^{\Theta/T}-1}\right]$$

which gives very close agreement with the experimental curve. At very low temperatures this reduces to

$$C_v = \frac{12\pi^4 \bar{R}}{5}\left(\frac{T}{\Theta}\right)^3$$

the Debye T^3 law.

Θ is known as the Debye temperature and equals hv_m/k where v_m is the upper limit of frequency of the atomic vibrations possible in the solid. At that temperature C_v is approximately 96·6 per cent. of the value of $3\bar{R}$.

A factor that has not been considered is the contribution to the specific heat of the electron gas in metals (the free valence electrons) which might be expected to contribute $\frac{3}{2}kT$ just as gas molecules should. However, because of Pauli's exclusion principle and the band structure of electron energies (Section 10.3) the electrons are, except at high temperatures, unable to absorb much energy and so make only a small contribution to the total.

Normally, for solids, the specific heat capacity is measured at constant pressure. This differs from that at constant volume by the energy necessary to compress the solid sufficiently to counter the effects of thermal expansion. It can be shown that

$$C_p - C_v = \alpha^2 VTK$$

where α is the coefficient of volumetric thermal expansion, V is the molar volume and K is the bulk modulus.

9.5 Thermal expansion

The curve of the variation of energy with internuclear spacing in a solid is of the form shown in Figure 9.3, c.f. Figure 5.5. At some temperature T_1 for which the energy is E_1, the atoms are vibrating so that the spacing r varies from A_1 to B_1 with a mean at r_1. At a higher temperature T_2, the vibration is between A_2 and B_2 with a mean position r_2. Since $r_2 > r_1$ due to the asymmetry of the curve, the solid expands. The amplitude of vibrations is of the order of 1/10 of the atomic spacing at ordinary temperatures. Melting occurs when the amplitude is about 12 per cent. The smaller the coefficient of thermal expansion, then the higher the melting point.

---

*For a derivation of this formula and others relating to specific heats, see J. de Lauray, 'The Theory of Specific Heats and Lattice Vibrations,' *Solid State Physics*, **2**, Academic Press, New York, 1956, p. 219.

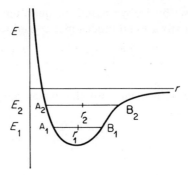

Figure 9.3. Variation of energy with internuclear spacing in a solid. E_1 and E_2 are average atomic energies at temperatures T_1 and T_2, respectively

9.6 Thermal equilibrium

It has already been shown (Section 6.5) that the molecules in a gas do not all have the same energy. There is continuously exchange of energy at every molecular collision and the velocities are distributed statistically according to the Maxwell–Boltzmann distribution.

When considering gas molecules, there is no restriction on the number of molecules that may have a particular energy value. For a system of electrons, however, there is a restriction that is expressed by Pauli's exclusion principle (see Section 4.3). Hence the statistical mechanical analysis of the two types of case will differ. An outline treatment is given in Appendix 8 which yields the following results.

For the case where there is no restriction on the numbers of particles per energy level, we get the Boltzmann distribution that the probability $p(E)$ of a particle having an energy value E is given by

$$p(E) = A \, e^{-E/kT}$$

For a system subject to Pauli's exclusion principle, we have the Fermi–Dirac distribution

$$p(E) = \frac{1}{e^{(E-E_0)/kT} + 1}$$

where E_0 is a constant energy value for any particular system. The significance of the value of E_0 will emerge in the next chapter.

Thermodynamics of Materials

9.7 Stable and metastable states

A system is said to be in a stable state of equilibrium when its free energy is a minimum. That is to say, work must be done (i.e. energy added) to disturb the system from this state, even slightly. Nevertheless many physical or chemical systems can exist more or less indefinitely in states which are not those of minimum free energy, but which require some addition of energy to change them to the lower energy stable state. The intermediate state is called a *metastable state*.

For example the stable condition for a mixture of hydrogen and oxygen at room temperature is the molecular form H_2O. If the gases are mixed at this temperature, they do not combine chemically but once a spark has passed they will react and in doing so give out heat energy, showing that the system has now attained a state of lower free energy and hence a more stable one.

9.8 Activation energy

In many physical and chemical changes, an atom or other particle can move from a metastable position to a more stable position by passing through an intermediate state of higher energy, as represented diagrammatically in Figure 9.4.

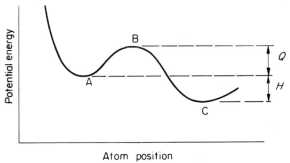

Figure 9.4.

An atom at position A can reach a more stable position C only by passing through the unstable position B. To do this it must receive the necessary additional energy to take it to the level B. This additional energy Q, i.e. the difference in the energy levels of positions A and B is the *activation energy*. In the hydrogen–oxygen mixture example cited above, the spark provides the activation energy. The final position C has a lower energy level so that in passing from B to C a total energy of $Q + H$ is released, giving a net energy release of H. This is termed the *heat of reaction* and may appear in any of the forms that energy can take: heat, light, sound, etc.

We have already seen that in a population of atoms or other particles there will be a distribution of energies among the particles. Those with energies equal to or greater than the activation energy will be able to cross the energy barrier from the metastable to the stable position.

The rate at which such a change can occur in a large population of such particles will depend on the magnitude of Q and also on the number of atoms that possess energy equal to or greater than Q at any instant.

When Q is small, other things being equal, the reaction can proceed faster than when Q is large.

We have already seen that for systems which are not subject to Pauli's exclusion principle the number of atoms which have an energy in the range E to $E + dE$ at a particular temperature T is given by the Boltzmann distribution

$$dN = A\,e^{-E/kT}\,dE$$

where k is Boltzmann's constant and A is a constant. The form of the distribution curve is shown in Figure 9.5.

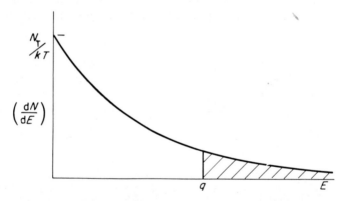

Figure 9.5. Maxwell–Boltzmann distribution law. Shaded area is proportional to number of atoms with energy greater than q

If N_T is the total number of atoms in the system, then

$$N_T = \int_0^\infty A\,e^{-E/kT}\,dE$$
$$= AkT$$

so that

$$A = N_T/kT$$

The number of atoms with energy greater than a particular value q is

$$N = \int_q^\infty \frac{N_T}{kT} e^{-E/kT} \, dE$$

$$= N_T e^{-q/kT}$$

If the energy is expressed as Q per kmol (i.e. $Q = N_0 q$), then the expression becomes

$$N = N_T e^{-Q/\bar{R}T}$$

Hence for a reaction with activation energy Q per kmol, the fraction of atoms with sufficient energy to overcome the barrier will be proportional to $e^{-Q/\bar{R}T}$. Of these a definite fraction (depending on the nature of the system) will move to the stable state so that the

$$\text{rate of reaction} = \mathscr{R} = B e^{-Q/\bar{R}T} = B e^{-q/kT}$$

where B is a constant. This is known as *Arrhenius' rate law*.

Writing this in the form

$$\log \mathscr{R} = \log B - \frac{Q}{\bar{R}T}$$

it will be seen that $\log \mathscr{R}$ varies linearly with $1/T$. If, for a particular reaction, experimental values of $\log \mathscr{R}$ when plotted against $1/T$ exhibit a straight line, then the reaction is a thermally-activated one and the slope of the line will be $-Q/\bar{R}$, from which the activation energy Q may be calculated.

9.9 Diffusion

The atoms of a crystal, in addition to vibrating about their mean positions, can also change positions with immediate neighbours or move into adjoining vacant sites, processes known as diffusion.* These processes differ from that of gaseous diffusion which is a transport phenomenon (see Section 6.7.3). In a homogeneous pure material, diffusion cannot be detected. If, however, some atoms of a radioactive isotope of the same element are introduced, these are in effect labelled atoms and diffusion can be observed. Atoms of a second kind can also diffuse into a pure metal, a process which can be observed and measured.

In such solid-state diffusion, an atom must pass through an intermediate position of higher energy. Its energy in the initial and final positions may be equal, so that it is not necessarily a passage from a metastable to a stable state, but from one stable state to another. However, it has to have the necessary

---

* For a review, see C. E. Birchenall, 'The Mechanism of Diffusion in the Solid State.' *Met. Rev.*, 3, 235 (1958).

activation energy to change places and so solid-state diffusion is a thermally activated process, proceeding at a faster rate at a higher temperature.

The rate at which diffusion can occur has been formulated into Fick's laws. If C is the concentration of the diffusing atoms at any point, and $\partial C/\partial x$ is the concentration gradient in the x direction, then the rate at which atoms cross unit area perpendicular to the x direction is given by Fick's first law

$$\frac{dm}{dt} = D\frac{\partial C}{\partial x}$$

where dm is the mass crossing unit area in time dt. Here D is the diffusion coefficient and has units of (length^2/time).

In general C is not constant at any one point, but varies with time as expressed by Fick's second law

$$\frac{\partial C}{\partial t} = \frac{\partial}{\partial x}\left(D\frac{\partial C}{\partial x}\right)$$

or if D is assumed to be constant

$$\frac{\partial C}{\partial t} = D\frac{\partial^2 C}{\partial x^2}$$

Solid-state diffusion is turned to practical use for many engineering purposes. In particular, its use in the manufacture of solid-state devices will be considered in Chapter 14.

Questions for Chapter 9

1. The pressure of a perfect gas is given by

$$p = \tfrac{1}{3}mn\overline{C^2}$$

where m is the mass of a molecule, n is the number of molecules per unit volume and $\overline{C^2}$ is the mean square velocity of the molecules. From this and the gas law $pV = RT$, determine the value of the specific heat capacity at constant volume C_v of a monatomic gas. Why does the value of C_v for a diatomic gas differ from this?

2. Estimate, from Dulong and Petit's law, the specific heat capacity at constant volume C_v for silver. By what percentage is the specific heat capacity at constant pressure C_p greater than C_v at 300 K?

    For silver, at 300 K, coefficient of linear thermal expansion = $1\cdot 86 \times 10^{-6}$ K^{-1}, density = $10\cdot 5 \times 10^3$ kg m^{-3}, bulk modulus = 100×10^9 N m^{-2}.

3. From considerations of Arrhenius' rate law, show that, if a system can exist in a metastable state with energy E_1 and a stable state with energy E_2

($E_1 > E_2$), then, under conditions of thermal equilibrium, the numbers of units (e.g. atoms, molecules) in the two states are given by

$$\frac{N_1}{N_2} = e^{-(E_1 - E_2)/kT}$$

For a hydrogen atom, the total energy of an electron is given by

$$E = -2 \cdot 18 \times 10^{-18}/n^2 \text{ J}$$

where n is the principal quantum number. In a flame at 3300 K containing 10^{20} hydrogen atoms, approximately how many atoms are in the first and in the second excited states? [ET]

4. A sample of cold-worked copper is found to be 50 per cent. recrystallized at the following combinations of heating times and temperatures:

207 °C	50 s
167 °C	7 min
127 °C	100 min
87 °C	2500 min

Show that these data are consistent with recrystallization being a thermally activated process and estimate the time for 50 per cent. recrystallization to take place at 27 °C. [MST]

5. The diffusion coefficient for silver diffusing into gold has the values $10^{-16 \cdot 1}$ and 10^{-12} m² s⁻¹ at 500 and 1000 °C, respectively. Assuming that diffusion follows Arrhenius' rate law, calculate the activation energy for diffusion per atom and per kmol.

6. If a concentration N_s of a solute atom is maintained in the surface layer of an initially solute-free solid the concentration N_d at a depth d below the surface after a time t due to the subsequent diffusion is given by

$$N_d = \frac{2N_s}{\sqrt{\pi}} \int_{x_0}^{\infty} e^{-x^2} dx$$

where $x_0 = d/2\sqrt{(Dt)}$ and D is the diffusion coefficient. Show that this expression satisfies Fick's second law.

CHAPTER TEN

Band Structure of Solids

10.1 Introduction

The picture of a metal as a regular array of positive ions with the valency electrons moving freely among them as an electron gas has been mentioned in Section 5.7. The free-electron theory, which is based on this concept, can explain some of the observed properties of metals, particularly electrical conductivity, but not all. The periodic nature of the electrical potential within the crystal lattice due to the regularly-spaced positive ions affects the electron waves. When the component of the wavelength resolved perpendicular to a set of atomic planes equals or nearly equals the interplanar spacing, a standing wave pattern is set up and the electron energy is modified. The result is that there is a range of energies which no electron can have. There are, therefore, *allowed and forbidden energy bands* associated with the material and these are termed the band structure. The qualitative treatment for the simpler metals is given first and a quantitative discussion follows later in this chapter. Non-metallic as well as metallic crystal structures show this property.

10.2 Energy bands

Imagine N atoms of an element arranged in a perfect crystal lattice, but with their interatomic distances many times larger than the normal value so that there is negligible interaction between the electron orbitals of different atoms. The quantum state for the crystal is then that for each atom duplicated N times. As the lattice constant is reduced, the wave functions of neighbouring atoms will overlap and the energy levels will be modified, as was discussed for two atoms in Section 4.1. The quantum states will no longer be restricted to individual atoms, but will extend over the whole crystal and the energy levels will be split so that each atomic energy level gives a band of levels or an *energy band*. Thus the $1s$ and $2s$ states might spread as in Figure 10.1. The $1s$ and $2s$ bands would each have N levels (each of which can take two electrons with

Band Structure of Solids

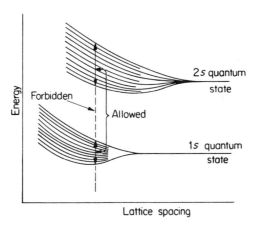

Figure 10.1. Typical spread of energy levels as lattice constant is decreased

opposite spin). The extent to which the band spreads is found to be independent of N, so that for a higher value of N (i.e. a larger crystal) the levels are more tightly packed.

Each electron in the crystal can only have an energy value which lies within one of the bands. If the energy bands do not overlap, as is the case in Figure 10.1, then no electron can have an energy value which would lie in the region between the bands, i.e. there will be a forbidden energy band.

The physical behaviour of the material depends very much upon the way in which the energy bands separate and overlap and also upon the extent to which they are occupied by electrons.

10.3 The band structure of the alkali metals

As stated previously, the atom of an alkali metal consists of a single valence electron outside an inner core of complete or stable shells. As the atoms move together, the energy levels of the valence electrons ($3s$ in sodium) split into a band as shown in Figure 10.2. There are N levels in the band, each capable of holding two electrons, which will have opposite spins, and there are N valence electrons so that only half of the available places are filled.

The value of the lattice spacing for which the total of the individual electron energies is a minimum will be the standard spacing of the crystal at 0 K. Obviously this will be when the lowest $N/2$ levels are filled, each with two electrons.

The energy levels of the inner electrons of the atoms also split when their corresponding wave functions overlap, but at the standard lattice spacing, the splitting is negligible.

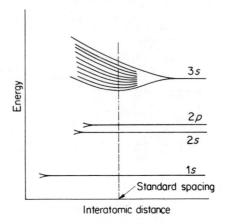

Figure 10.2. Spread of energy bands of sodium—diagrammatic only

10.4 The band structure of the alkaline earth metals

The alkaline earth metals (Mg, Ca, etc.) have two valence electrons per atom and so it would appear that in a crystal the band corresponding to the s valence level would be full. However, the possible excited states must also be considered and the energy levels of these also spread into bands. Thus in magnesium, the 3p and 3d bands both spread and overlap the 3s band as shown on Figure 10.3.

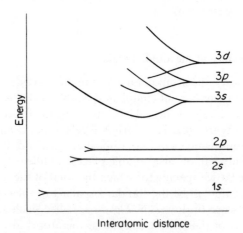

Figure 10.3. Diagrammatic representation of energy levels in magnesium

Band Structure of Solids

Each 3p level in an atom has six quantum states, so that the 3p band can hold 6N electrons. Similarly, the 3d band can hold 10N electrons. Hence in the ground state, when the 2N valence electrons will occupy the lowest energy levels, the majority will be in the 3s state and the remainder in 3p and 3d states.

10.5 Occupancy of energy levels

At 0 K, the electrons will occupy the lowest available energy levels (i.e. in sodium the lower half of the 3s band will be filled and the upper half empty). At higher temperatures, some electrons will receive energy by thermal excitation and be moved to higher energy states. Also, once there are some vacant lower energy states, electrons can fall into them giving up the excess energy. A state of dynamic equilibrium will be reached, the state being dependent upon the temperature.

The occupancy of energy states by electrons will follow Fermi–Dirac statistics (see Section 9.6) and the distribution of the electrons among the possible energy states can be shown by a *Fermi distribution* curve. This shows the probability $p(E)$ of any quantum state of energy E being occupied. If the probability is one, the state is always full; if it is zero, the state is always empty and a fractional probability means that it is occupied for a part of the time. At 0 K, the electrons occupy fully the lowest states up to some energy value E_{max} and states with energies above that value are empty. E_{max} is equal to the arbitrary constant E_0 in the Fermi–Dirac distribution function. It is known as the *Fermi level* in this case. A more exact definition will appear later. The Fermi distribution curve for this temperature is shown in Figure 10.4(a). For any but the smallest piece

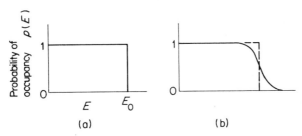

Figure 10.4. Fermi distribution curves at (a) 0 K, (b) $T(>0\,\text{K})$

of material, N is large so that the allowed energy levels are very closely spaced and a continuous curve for $p(E)$ can be drawn. E_{max} for most metals is of the order of 5 eV.

At temperatures higher than 0 K, electrons may absorb energy by transfer of thermal energy from atoms and molecules, but energy can be absorbed only if there is a vacant state of higher energy into which the electron can move. The mean thermal energy which can be transferred to an electron in one step is of the order of kT (see Section 9.4) which at room temperature (18 °C) is about 0·025 eV. This is very much smaller than E_{max} and so only those electrons with energies which differ from E_{max} by about kT or less can be excited to higher levels. The excited electrons will, in time, fall back to the lower vacant levels so that a state of dynamic equilibrium is established, being a balance between the two continuous processes.

In a metal such as sodium, the fraction of the electrons involved is of the order of kT/E_{max} and the average energy absorbed by each is of the order of kT, so that the total thermal energy is of the order $(kT)^2/E_{max}$. The electronic contribution to the specific heat will be the change of energy for a temperature change of 1 K, i.e. $2k^2T/E_{max}$ per atom. At 300 K this is approximately $k/100$. The specific heat of solids approaches $3k$ per atom at temperatures above the Debye temperature (see Section 9.4) and so the electronic contribution is small—about $\frac{1}{300}$ at room temperature and $\frac{1}{100}$ at 600 °C. However, at temperatures near 0 K, when the contribution of lattice vibrations is small, the electronic contribution is an appreciable fraction of the total. Some metals have a concentration of energy levels near the top of the occupied portion of the conductor band (see Section 10.9). In these cases, the electronic contribution to the specific heat is considerably higher.

When $T > 0$, then the Fermi distribution curve will take a shape like that given in Figure 10.4(b).

10.6 Distribution of available energy states

From the wave-mechanical viewpoint, the valence electrons of a metal are moving in the potential field of the positive ions which are arranged in the regular pattern of a crystal lattice. This field will be the sum of a series of potential wells and everywhere the potential will be negative. As a first approximation, this can be considered as being smoothed out to a uniform value, so that the electrons are moving in a three-dimensional potential box. The wave-mechanical solution for the motion of an electron in such a box is given in Section 3.4, from which it is seen that the electron can only have one of certain permitted energies which are given by equation (3.12). It was also shown that the number of stationary states with energy less than E is

$$\frac{\pi}{6} \cdot \left(\frac{8mE}{h^2}\right)^{3/2} V$$

where V is the volume of the box.

Band Structure of Solids

The number of electrons that can be accommodated in all states up to E_{max} is equal to twice the number of states, i.e.

$$N = \frac{\pi}{3}\left(\frac{8mE_{max}}{h^2}\right)^{3/2} V$$

and hence

$$E_{max} = \frac{h^2}{8m}\left(\frac{3N}{\pi V}\right)^{2/3}$$

E_{max} depends upon N/V which is the number of free electrons per unit volume. This is independent of the size of the piece of metal, and so E_{max} is constant for any one metal.

Example. Calculate the value of E_{max} for silver.

Silver has a face-centred cubic lattice so that there are four atoms per unit cell and

$$V/N = a^3/4$$

where a is the lattice constant ($=4\cdot086$ Å). Hence

$$E_{max} = \frac{(6\cdot625 \times 10^{-34})^2}{8 \times 9\cdot1083 \times 10^{-31}}\left(\frac{3 \times 4}{\pi(4\cdot086 \times 10^{-10})^3}\right)^{2/3} \text{ J}$$

$$= 8\cdot82 \times 10^{-19} \text{ J}$$

$$= \frac{8\cdot82 \times 10^{-19}}{1\cdot602 \times 10^{-19}} \text{ eV}$$

$$= 5\cdot5 \text{ eV}$$

As seen from the above example, the value of E_{max} for silver is 5·5 eV. The value for other metals will be of the same order of magnitude.

If $N(E)$ is defined as the density of available states, i.e. $N(E)\,dE$ is the number of possible states in unit volume with energies between E and $E + dE$, then $N(E)$ is the differential with respect to E of the total number of states up to an energy E, i.e.

$$N(E) = 4\pi\left(\frac{2m}{h^2}\right)^{3/2} E^{1/2} \tag{10.1}$$

The curve of $N(E)$ against E is a parabola as shown in Figure 10.5(a). At 0 K, all states with energies up to the Fermi level are occupied and all states with higher energies are vacant, so that the distribution of occupied states is as shown

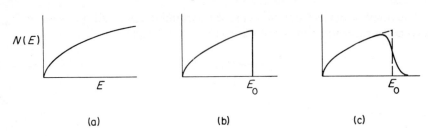

Figure 10.5. (a) Density of available states as a function of energy, (b) states occupied at 0 K, (c) states occupied at $T (>0\,\text{K})$

in Figure 10.5(b). At a temperature T, the curve of $N(E)$ has to be multiplied by the Fermi–Dirac distribution function and gives a distribution of occupied states as in Figure 10.5(c). Now the Fermi–Dirac distribution function $p(E)$ for $E_0 + \delta E$ is equal to $(1 - p(E))$ for $E_0 - \delta E$, that is the probability of a state δE above the Fermi level E_0 being occupied is equal to the probability of the state δE below the Fermi level being vacant. But the value of $N(E)$ is greater for $E_0 + \delta E$ than it is for $E_0 - \delta E$, so that if the total number of occupied states is to be constant, E_0 must decrease slightly as the temperature increases.

The parabolic shape for the $N(E)$ versus E curve discussed above could continue to higher energies only as far as it would not be affected by the approach to a forbidden band. This condition is satisfied for the monovalent alkali metals and silver up to the energy levels needed to accommodate the valence electrons. At the top of the allowed band, other factors come into play so that the value of $N(E)$ decreases with increasing E until it reaches the value 0 at the top of the allowed band. The factors that cause this variation of $N(E)$ are discussed in the following sections.

10.7 The effect of the periodic field

The treatment given in Section 3.4 shows that the only possible electron states in a box with dimensions a, b, c are those with wave functions given by

$$\psi = A \sin \frac{\pi n_x x}{a} \sin \frac{\pi n_y y}{b} \sin \frac{\pi n_z z}{c}$$

where n_x, n_y, n_z are integers, i.e. the electron density would vary sinusoidally with position. As stated in that section, this is the standing wave due to reflection of the travelling electron wave

$$\psi = A\, e^{i(k_x x + k_y y + k_z z - 2\pi vt)}$$

at the sides of the box.

Band Structure of Solids

The periodic field introduces regularly spaced boundaries of the type considered in Section 3.1 with $E > V_1$, i.e. there will be partial reflection at each boundary. The problem will be considered firstly as a one-dimensional case and then the treatment will be extended to a consideration of the three-dimensional crystal.

Consider electron waves travelling in the x direction in a potential field which varies in some manner as shown in Figure 10.6. All the atoms in any one plane

Figure 10.6. Possible periodic variations of potential **(a)** in line through centres of atoms, **(b)** in parallel line not through centres of atoms

perpendicular to the x direction will act as partially-reflecting centres and send out wavelets of the type similar to Huygens wavelets in optics. These will interfere and cancel out except in the x direction where they will give a forward and a reflected plane wave travelling in the x and $-x$ directions, respectively.

Let the incident wave be

$$\psi = A\,e^{i(kx - 2\pi vt)}$$

and let the wave reflected from one plane of atoms be

$$\psi = A_1\,e^{-i(kx + 2\pi vt)}$$

Now as each plane of atoms reflects a wave, these reflected waves will interfere with one another. In general, for an arbitrary value of k they will not be in phase and so the interference will be destructive and cancel out. In other words, the initial electron wave will traverse the crystal without reflection and will behave as a free electron within the crystal.

If, however, the waves reflected from successive planes of atoms are in phase, i.e. the value of k is such that the path difference $2a$ equals an integral number of wavelengths, then the successive reflections combine to form a strong reflected wave. The condition for this to occur is that

$$k = \pm \pi n / a$$

where n takes any integral value.

When k has a value which satisfies this equation, the wave function can no longer be represented, even approximately, by e^{ikx} which represents an electron travelling in one direction. The correct functions must include equally strong incident and reflected waves, and these are

$$e^{ikx} + e^{-ikx} (= 2\cos kx)$$
$$e^{ikx} - e^{-ikx} (= 2i\sin kx)$$

which represent standing waves. The cosine function has its nodes where the sine function has its maxima, and vice versa. One will give the maximum electron density in the troughs of the potential V (Figure 10.6) which corresponds to lower energy than for a free electron and one will give the maximum density at the potential peaks which corresponds to a higher energy.

Hence the effect of the periodic potential variations is that at each critical value of k the free electron energy level is split into two levels. Between these energy levels there is no allowed state of motion for the electron so that there is a forbidden energy gap, the width of which depends upon the amplitude of the periodic part of V.

For a crystal with no periodic potential,

$$E = k^2 h^2 / 8\pi^2 m$$

i.e. a parabolic relationship as shown in Figure 10.7(a). With a periodic field,

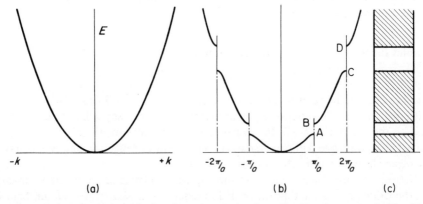

Figure 10.7. E–k relationship for electrons moving in x direction in a crystal: **(a)** with constant potential, **(b)** with periodic potential, **(c)** energy bands (forbidden bands are not shaded)

the E–k curve flattens off at the critical values of k producing discontinuities A to B, C to D, etc. as shown in Figure 10.7(b). The energies within these ranges are known as forbidden bands, Figure 10.7(c). Within each allowed band, the energies are closely spaced and form a quasi-continuous spectrum.

Band Structure of Solids

10.8 The three-dimensional periodic field

In a simple cubic lattice, of lattice constant a, consider an electron wave travelling in a direction such that the normal AB to the wave front lies in the x–y plane and makes an angle θ to the y–z plane as shown in Figure 10.8.

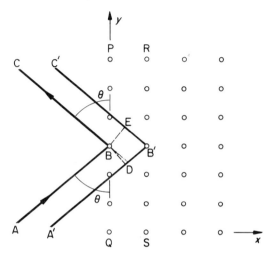

Figure 10.8. Reflection of electron waves by planes of atoms

The wavelets reflected from the line of atoms PQ will be in phase and give a reflected wave front with a normal in the direction BC, where BC also makes an angle θ with PQ. Each layer of atoms parallel to PQ will give a reflected wave front and, as in the previous section, these wave fronts will cancel one another unless they are in phase. If ABC and A'B'C' are two sets of normals to the wave fronts corresponding to reflections from successive layers PQ and RS, then B and D are in phase in the incident wave. B and E are corresponding points in the reflected waves and the path difference is DB' + B'E'. The successive reflections are in phase when this path difference is an integral number of wavelengths, i.e. when

$$2a \sin \theta = n\lambda$$

which is also Bragg's law for the reflection of X-rays from planes of atoms.

In considering electron waves travelling in different directions, it is convenient to specify the electron wave by the wave vector **k** (introduced in Section 3.4), which has a direction normal to the wave front and magnitude k equal to

$2\pi/\lambda$. Then

$$k = \pm \frac{n\pi}{a \sin \theta}$$

for the reflection of the electron waves.

Now the wave vector can be resolved into components k_x, k_y and k_z parallel to the x, y and z axes, respectively. Of these,

$$k_x = k \sin \theta$$

so that the positions of the discontinuities in the E–k curve for reflections from the (100) planes are given by

$$k_x = \pm \frac{n\pi}{a}$$

Similarly for reflections from the (010) and (001) planes the positions of the discontinuities will be given by

$$k_y = \pm \frac{n\pi}{a}$$

and

$$k_z = \pm \frac{n\pi}{a}$$

respectively.

These results are conveniently shown on a diagram using k space (see Section 3.4). Thus for a two-dimensional case, considering the x–y plane, the first discontinuity occurs when either

$$k_x = \pm \frac{\pi}{a}$$

or

$$k_y = \pm \frac{\pi}{a}$$

and is shown by the square ABCD of Figure 10.9.

Discontinuities will also occur for higher order reflections from these planes and for reflections from other planes. Thus the reflection from the {110} planes would occur when

$$k = \frac{n\pi}{(a/\sqrt{2}) \sin \phi}$$

Band Structure of Solids

for waves with k at an angle ϕ to the $\{110\}$ planes; that is for

$$\pm k_x \pm k_y = \sqrt{2}\frac{n\pi}{a}$$

which correspond to lines at 45° to either axis and passing through points A, B, C and D on Figure 10.9.

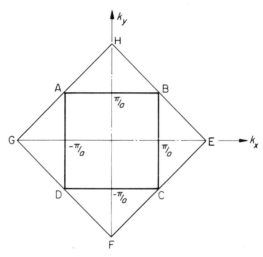

Figure 10.9. First and second Brillouin zones for two-dimensional square lattice

Thus k space is divided up into regions by surfaces which correspond to k values at which there are discontinuities in the E–k curves. These regions are known as *Brillouin zones*.

For the two-dimensional case considered, the square ABCD in Figure 10.9 is the boundary of the first Brillouin zone and the inclined square EFGH is the boundary of the second Brillouin zone. The boundaries of the zones are parallel to the planes in the crystal that give rise to the zones and the perpendicular distance of a boundary from the origin is $n\pi/d$ where d is the spacing of the reflecting planes.

The E–k relationship for electron waves with wave vectors in any direction will be of the form shown in Figure 10.7(b), but for different directions the gaps will correspond to different values of E. Thus for the x direction, the E–k curve would be as in Figure 10.10(a) and for the [110] direction it would be as in Figure 10.10(b). The forbidden bands have different mean energy levels and may or may not overlap depending upon their widths.

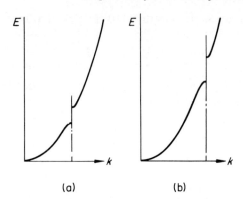

Figure 10.10. Variation of E with k: **(a)** along [100] direction, **(b)** along [110] direction

The variation of E with k can also be shown as a series of energy contours in k space, each contour connecting those values of k which have equal energy, as in Figure 10.11. For values of **k** which are far from the critical values at the

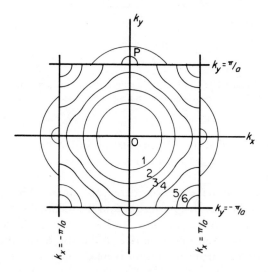

Figure 10.11. First Brillouin zone for two-dimensional square lattice showing equal energy curves

boundaries, $E \propto k^2$ so that the contours are circles concentric with the origin. Nearer the zone boundaries, however, the curves will bulge towards the boundaries because near each critical k value, E rises less rapidly than the parabolic

Band Structure of Solids

relationship would infer. The higher energy contours would be confined to the corners of the zone. Because of the discontinuity in the E–k relationship, the contours do not continue across the zone boundary. The lowest energy in the second zone will occur at a point such as P. This energy may be higher or lower than the highest energy of the first zone, which occurs at a corner. If it is lower, then some electrons will occupy states represented by points in the second zone before all the states for the first zone are filled (Figure 10.12(a)), i.e. the energy bands overlap. If it is higher, then all the states represented by points in the first zone must be filled before any electrons occupy a state in the second zone (Figure 10.12(b)).

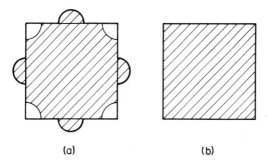

Figure 10.12. Brillouin zones with (**a**) overlapping, (**b**) non-overlapping energy levels

10.9 The Fermi surface

For a complete treatment of metals, the z direction must also be considered, so that the Brillouin zones become polyhedra. For a simple cubic lattice, the first Brillouin zone would be a cube of side $2\pi/a$. For a face-centred cubic lattice, the first Brillouin zone (Figure 10.13) is due to reflections from the $\{200\}$ and $\{111\}$ planes.

Figure 10.13. Brillouin zone for face-centred cubic lattice

In the one-dimensional problem, the possible values of k were $2\pi n/L$ where L was the length of the potential box, so that successive values of k differed by $2\pi/L$, or each k value was allotted a length $2\pi/L$ of the x axis in Figure 10.7(a).

In three dimensions, the volume of k space associated with each quantum state (one value of k is defined by the three quantum numbers n_x, n_y and n_z) is $8\pi^3/V$ where V is the volume of the crystal.

For a simple cubic lattice, the volume of the first Brillouin zone is $8\pi^3/a^3$. Hence the number of quantum states in the zone is $(8\pi^3/a^3)/(8\pi^3/V)$, i.e. V/a^3, which is equal to the number of lattice points in the crystal. Each state can take two electrons with opposite spins so that the zone can hold twice as many electrons as there are lattice points. For face-centred and body-centred cubic lattices, the same result holds. For hexagonal close-packed lattices, the actual number depends upon the axial ratio of the unit cell and is usually slightly less than two.

The contours of equal energy are surfaces in three dimensions. That one which corresponds to the Fermi level at 0 K is known as the *Fermi surface*.

In sodium, which has a body-centred cubic structure and one valence electron per atom, the first Brillouin zone is half full so that the Fermi surface is approximately spherical. In copper, the surface is distorted from a spherical shape and touches some of the surfaces of the Brillouin zone as shown in Figure 10.14.

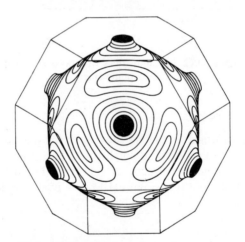

Figure 10.14. Fermi surface of copper as determined by Pippard (*Phil. Trans. Roy. Soc. A*, **250**, 325, 1957). The curves are contours of constant distance from the centre. The Fermi surface touches the Brillouin zone boundary in the vicinity of the [111] directions, the black areas denoting the regions of contact. (By permission of the Royal Society)

In the case of the alkaline earth metals, there are two valence electrons per atom so that there are just enough to fill the first zone in face-centred cubic

Band Structure of Solids

calcium and strontium and in body-centred barium and more than enough to fill the first zone in hexagonal close-packed magnesium. However, the energy band corresponding to the second Brillouin zone overlaps that corresponding to the first zone in each case. The Fermi surface for each of these metals will therefore be a more complicated shape.

The volume of the second Brillouin zone for a simple cubic lattice would be $24\pi^3/a^3$ so that the number of quantum states in the zone is $3V/a^3$ or three per lattice point. Hence the zone could hold six electrons per lattice point, corresponding to the number of p electrons in a shell of an atom. The spread of energies for the p and d levels are usually less than that of the s level and, because of the greater number of electrons per atom that these sub-levels can hold, the $N(E)$ values will be higher. The shape of the $3d$ band which overlaps the $4s$ band is shown in Figure 10.15. The tall but narrow $3d$ level is important in consideration of the ferromagnetic materials which will be discussed in Chapter 15.

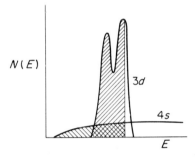

Figure 10.15. Overlapping $3d$ and $4s$ energy bands (diagrammatic only). The shaded areas represent the occupance of the bands in iron at 0 K. The $3d$ band holds 7·4 electrons and the $4s$ band holds 0·6 electrons per atom

10.10 The relation of energy bands and conductivity

It will be shown in the next chapter that a material can conduct electricity only if there are immediately adjacent vacant energy levels into which the highest energy electrons can move. This condition is satisfied by any material which has a partially full energy band.

When there are just sufficient electrons to fill an allowed band which does not overlap the next allowed band, then the material will not conduct electricity at 0 K. If the gap between these energy bands is very wide, so that electrons cannot receive enough energy by thermal or other means to cross the gap, then the material is an *insulator*. If, however, the gap is small, then either thermal energy or a sufficiently high electrical field may cause some electrons to cross the gap into the upper band. There they will have immediately adjacent vacant levels and so the material can conduct electricity. Such materials are known as *semiconductors* and are considered further in Chapter 13. The nature of the energy bands in these three classes of materials is shown schematically in Figure 10.16.

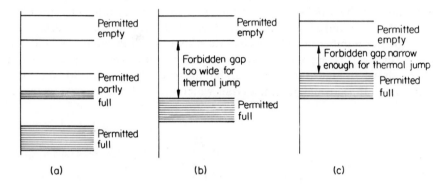

Figure 10.16. Typical energy band and electron distribution in (a) conductor, (b) insulator and (c) semiconductor

Questions for Chapter 10

1. Calculate the value of the energy of the Fermi level E_0 for sodium at 0 K.

2. Show that the average kinetic energy at 0 K of all the valency electrons in sodium is $\frac{3}{5}E_0$.

3. Show that for a metal with a parabolic $N(E)$ versus E relationship, the contribution to the specific heat capacity of the free electrons is of the order of 100 J kmol^{-1} at 27 °C.

4. By considering two-dimensional cases for the (100) and the (110) planes of a face-centred cubic lattice, deduce the form of the first Brillouin zone. What is the number of quantum states per lattice point within the zone?

5. By similar consideration of (100) and ($2\bar{1}0$) planes of a hexagonal close-packed lattice with $c/a = \frac{5}{3}$ deduce the form of the first Brillouin zone.

6. (a) By considering Schrödinger's equation, find and sketch the relation between energy E and k for a free electron where $k = 2\pi/\lambda$, λ being the de Broglie wavelength.

Band Structure of Solids

(b) By reference to a potential barrier, show briefly and in principle only how it is that quantum mechanics predicts that an electron may travel freely through the structure of Figure 10.17 without getting trapped in the electric potential wells, when its energy is less than the height of the barrier.

Figure 10.17

(c) Figure 10.18 shows the result of a calculation for E as a function of k for an electron moving in a periodic potential of the type shown in Figure 10.17. The energy E is taken with respect to some arbitrary level.

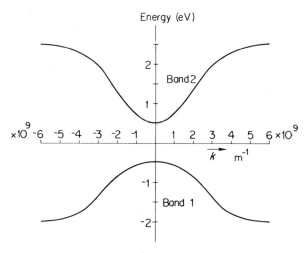

Figure 10.18

Show that the electrons at low enough energies in band 2 behave as free electrons but with an effective mass m^*. Using the data of Figure 10.18 estimate the effective mass in terms of the free-electron mass. [EST]

CHAPTER ELEVEN

Conducting Materials

11.1 Electrical conductivity

In a metal, in the absence of external electrical fields, there will be electrons moving in each direction in equal numbers, but this dynamic equilibrium will be upset in the presence of a field. It is convenient to discuss the problem in terms of the wave vector **k** of the electron. This was defined in Section 3.4 as $2\pi/\lambda$ and hence according to de Broglie's hypothesis equals $2\pi mv/h$, i.e. it is proportional in magnitude to the speed of the electron and is in the direction of the electron velocity. An electron with a higher **k** value will have a higher energy E.

Consider a one-dimensional model of **k** space. The Fermi distribution in terms of the wave vector would at 0 K be as shown in Figure 11.1 by the full line. All

Figure 11.1. Occupance of electron states in one-dimensional lattice at 0 K in absence (full line) and presence (broken line) of electric field

values of k from $-k_m$ to $+k_m$ would be occupied, k_m being the value of k which corresponds to the energy E_0 of the Fermi level. For each electron moving in one direction, there would be another electron moving in the opposite direction with equal speed, so that the net current would be zero. Applying an electric field $\mathscr{E}$ would tend to accelerate all the electrons in one direction. This would change the k value of each electron and some electrons would have k values in excess of k_m. This can occur only if there are vacant k values above k_m which can be occupied. If all the states in the energy band are not full, then this dis-

Conducting Materials

placement of k values will occur, giving the distribution shown by the broken line in Figure 11.1. Most electrons will still be paired, but those near $+k_m$ are not, so that there is a net drift of electrons, i.e. an electrical current.

If the energy band were full and did not overlap the next band, then there would be no k values adjacent to and just greater than $+k_m$ so that the displacement could not occur and the material would not conduct electricity.

11.2 Mobility and resistivity

Under the influence of an applied electric field, each electron in a solid which is not bound to a specific atom would tend to be accelerated in a direction opposite to that of the electric field. Only those electrons for which there are vacant energy states of slightly greater energy can, however, be accelerated. In any reasonable-sized crystal, the energy states are so closely packed that, if conductivity is possible, the states can be assumed to be a continuous distribution for analysis purposes.

Continuous acceleration would imply a conductivity increasing continuously with time. However, the conductivity of a material is not time dependent but, for other variables remaining constant, is itself a constant. Therefore, the average velocity acquired against the direction of the field must be constant.

This can be explained by assuming that the electrons undergo collisions, at each of which there will be a redistribution of the extra energy which the electron has acquired. In a perfect crystal structure, the electrons move in a potential field which is regularly periodic in three dimensions. Except for certain velocities, which correspond to the forbidden energy bands, the electrons would not be scattered internally. Hence the collisions discussed above can only occur due to the deviations from the ideally perfect lattice. The types of deviation or imperfection are discussed below.

Even without the externally applied field, the electrons would suffer such collisions and these will be distributed randomly in time.

Hence the probability that an electron undergoing such random motion shall collide in a time interval δt is constant no matter what time has elapsed since it last collided. Let this probability be $\delta t/\tau$, where τ is a constant. For a large density of imperfections, τ will be smaller and the conductivity less.

Consider a population of electrons and let $n(t)$ be the number which have been in motion for a time t without a collision. Then in the interval from t to $t + dt$, the probable number of those which will collide is $n(t)\,dt/\tau$. But this is the change in $n(t)$ so that

$$dn = -n(t)\,dt/\tau$$

which is a differential equation giving

$$n(t) = n_0\,e^{-t/\tau} \qquad (11.1)$$

where n_0 is the value of $n(t)$ for $t = 0$. Then the number colliding between t and $t + dt$ is $n_0 e^{-t/\tau} dt/\tau$.

Alternatively, the times between the collisions of a single electron will be distributed in a similar manner.

Under the influence of an applied electric field each electron will experience an acceleration

$$\alpha = e\mathscr{E}/m$$

due to the field, where e and m are the electron charge and mass, respectively.*

If one electron travels for a time t between two successive collisions, then the extra distance travelled due to the acceleration is $\frac{1}{2}\alpha t^2$. After this electron has made n collisions the distance travelled in the direction of the acceleration is

$$\tfrac{1}{2}\alpha(t_1^2 + t_2^2 + t_3^2 + \cdots + t_n^2)$$

where $t_1, t_2, t_3, \ldots, t_n$ are the free times between collisions. These times will be distributed statistically according to equation (11.1) provided that the velocity acquired due to the acceleration is always small compared with the mean random velocity. If n is large, then

$$t_1^2 + t_2^2 + t_3^2 + \cdots + t_n^2 = \int_0^\infty t^2 n\, e^{-t/\tau}\, dt/\tau$$

$$= 2n\tau^2$$

so that the distance travelled is $\alpha n\tau^2$.

The time taken for these n collisions to occur is

$$t_1 + t_2 + t_3 + \cdots + t_n = \int_0^\infty tn\, e^{-t/\tau}\, dt/\tau$$

$$= n\tau$$

Therefore τ is the average time between collisions and is known as the *relaxation time*.

---

* More exactly, the effective mass m^* should be used. The relationship $d^2E/dk^2 = h^2/4\pi^2 m$ only applies when the $E - \mathbf{k}$ relation is parabolic. Near a forbidden band, the curvature of the $E - \mathbf{k}$ curve changes and we get

$$m^* = \frac{h^2}{4\pi^2}\frac{d^2E}{dk^2}$$

and the acceleration is given by

$$\alpha = e\mathscr{E}/m^*$$

The effective mass equals m away from the forbidden bands. Near the top of the allowed band the curvature changes sign so that d^2E/dk^2 and hence m^* become negative, i.e. an electron may be accelerated in the direction of the electric field. This is really due to diffraction effects which dominate when the forbidden band is approached.

Conducting Materials

Hence the drift velocity is

$$v = \frac{\alpha n \tau^2}{n\tau}$$

$$= \alpha \tau$$

$$= \frac{e\mathscr{E}\tau}{m}$$

or

$$v = \mu_n \mathscr{E}$$

where $\mu_n (= e\tau/m)$ is the *mobility* of the conducting electrons. As long as the drift velocity is small compared with the thermal velocities of the electrons, then the current density, which equals the charge crossing unit area in unit time, is

$$J = env$$

$$= en\mu_n \mathscr{E}$$

where n is the number of conducting electrons per unit volume.

The conductivity σ of a material is defined by Ohm's law

$$J = \sigma \mathscr{E}$$

so that the conductivity due to the conducting electrons is

$$\sigma = en\mu_n$$

Or, the resistivity ρ, which is the reciprocal of the conductivity, is given by

$$\rho = \frac{1}{en\mu_n}$$

11.3 Factors affecting resistivity

The imperfections that can cause increased scattering of conducting particles are either chemical, such as foreign substitutional and interstitial atoms, or geometrical, such as vacancies and dislocations, or thermal caused by the thermal vibrations of atoms about their equilibrium positions.

The first of these is typified by the increase of resistivity of a metal as the alloy content is increased (Figure 11.2), and the second by the increase of resistivity due to work hardening (Figure 11.3), which introduces a high dislocation density.

The scattering due to thermal vibrations increases as the temperature is raised. In the case of a metal, where the number of carriers is relatively independent of temperature, the scattering due to thermal vibrations is the only

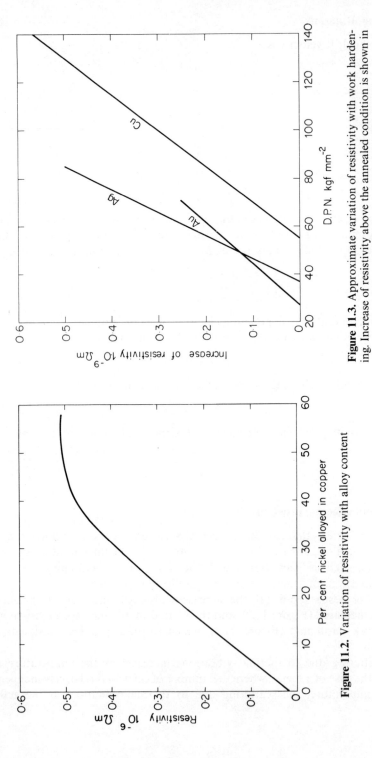

Figure 11.3. Approximate variation of resistivity with work hardening. Increase of resistivity above the annealed condition is shown in relation to the Diamond Pyramid Hardness

Figure 11.2. Variation of resistivity with alloy content

Conducting Materials

temperature-sensitive variable. The total resistivity is the sum of two terms, one ρ_0 due to the fixed defects and the other ρ_T due to the thermal vibrations.

$$\rho = \rho_0 + \rho_T$$

This equation is known as *Mattheison's rule*. At temperatures well above the Debye temperature, ρ_T is approximately proportional to the absolute temperature. That is if ρ_r is the room temperature resistivity, then

$$\rho = \rho_r(1 + \alpha T)$$

approximately, where T is the temperature above room temperature. The *temperature coefficient of resistivity*, α, is about 0.004 K^{-1} for pure metals and generally lower for alloys. Some examples of variation of resistivity with temperature are shown in Figure 11.4.

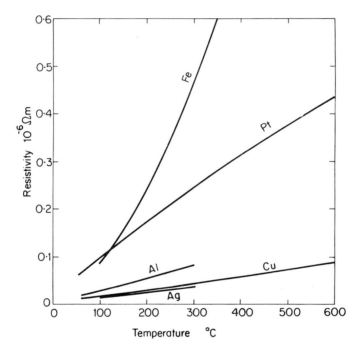

Figure 11.4. Variation of resistivity with temperature

11.4 Thermal conductivity

In a metal, the valence electrons with energies near the Fermi level also act as conductors of thermal energy. In ionic and covalent materials, thermal conductivity is due solely to the transfer of atomic vibrations by elastic interactions, i.e. heat is transferred as elastic waves or *phonons*, which are also reflected and scattered at lattice irregularities. Because the mean electron velocity is two or more orders of magnitude greater than that of phonons, then when conducting electrons are present, their contribution to thermal conductivity is much higher than that due to waves. Hence metals have much higher thermal conductivities than materials which are electrical insulators.

Also in a metal, higher electrical conductivity is associated with higher thermal conductivity, the same number of conducting electrons being responsible for each. At higher temperatures, each conducting electron will carry, on average, higher thermal energy so that the ratio of thermal conductivity K to electrical conductivity σ is proportional to absolute temperature. This is summarized in the experimentally-determined *Wiedmann–Franz law*

$$\frac{K}{\sigma T} = L$$

where L is constant for each metal having values between 2 and $3 \times 10^{-8}\,\text{V}^2\,\text{K}^{-2}$. The theoretical value of L is $2\cdot 45 \times 10^{-8}\,\text{V}^2\,\text{K}^{-2}$.

11.5 Work function

It has already been said that the valence electrons in a metal are not bound to any particular atoms, but are free to move over the whole extent of the metal. In the interior, the potential will vary in a periodic manner, the exact nature of which is not important here and at the surface there is a potential energy barrier. This is not a step function but varies in some manner such as is shown in Figure 11.5. As the distance from the surface increases, the potential energy curve tends asymptotically to its value at infinity.

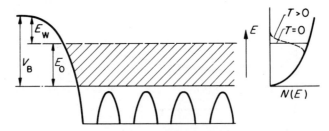

Figure 11.5. Variation of potential near surface of metal. Fermi curve shows electron distribution at 0 K and at a higher temperature

Let the height of the potential barrier above the bottom of the conducting band be V_B and the Fermi level be E_0 above the same reference energy. The

Conducting Materials

difference of these two quantities

$$E_W = V_B - E_0$$

is known as the *work function*. It is the minimum energy which must be given to an electron if it is to escape from the solid at the absolute zero of temperature.

One manner in which this energy can be given to an electron is by receiving it from a photon, as described in Section 2.3. The photoelectric work function mentioned there is E_W.

At temperatures above absolute zero, some electrons will have extra thermal energy and so require to be given energy less than E_W to escape. This must be allowed for in attempting to derive a value of E_W from photoelectric measurements.

11.6 Thermionic emission

At temperatures above absolute zero, the energies of the electrons in a metal will be distributed according to the Fermi distribution function. Those with energies which exceed that of the potential barrier at the surface can escape from the metal. As the temperature is increased, a higher proportion will have sufficient energy and the number escaping will increase. This escape of electrons which have sufficient thermal energy is known as *thermionic emission*.

Thermionic emission can be investigated by heating a cathode in vacuo and having near it an anode which is maintained at a sufficiently positive voltage relative to the cathode to attract all the emitted electrons. The current flowing to the anode from the potential source is then equal to the total emission. It is shown in Appendix 9 that the current density of the thermally emitted electrons is given by the *Richardson–Dushman* equation

$$J = A_0 T^2 e^{-E_W/kT} \qquad (11.2)$$

A_0 is a universal constant with the theoretical value

$$A_0 = 1 \cdot 2 \times 10^6 \text{ A m}^{-2} \text{ K}^{-2}$$

From this equation it can be seen that

$$\log_e \frac{J}{T^2} = \log_e A_0 - \frac{E_W}{k} \frac{1}{T}$$

so that a plot of $\log J/T^2$ versus $1/T$ should give a straight line from which values of A_0 and E_W can be determined.

As examples, the values for nickel and tungsten are

	A_0	E_W
Ni	$30 \times 10^4 \, \text{A m}^{-2} \text{K}^{-2}$	4·61 eV
W	30×10^4	4·52

The experimental values of A_0 are generally smaller than the theoretical value.

The exponential term in the Richardson–Dushman equation is the most significant one when the temperature changes. For example in tungsten the emission increases by a factor of 2×10^5 for a temperature increase from 1000 to 1500 °C.

High thermionic emission can be achieved by using higher temperatures or by using materials with lower work functions. Tungsten with a melting point of 3380 °C is most useful when higher temperatures are to be used.

By adsorbing a layer of suitable foreign substance on to the surface, the work function can be lowered and a higher emission achieved at a lower temperature. Thorium, for example, lowers the work function of tungsten to 2·6 eV and this increases the emission by a factor of 4×10^7 at 1000 °C. The thorium is added as thorium dioxide ThO_2 which reduces grain growth and prevents premature failure of the tungsten due to embrittlement and loss of mechanical strength. Some of the thorium dioxide is reduced to metallic thorium and migrates to the surface where it has the effect described above.

Surfaces made from a mixture of the oxides of the alkaline earth metals are found to be very good for thermionic cathodes. They are made by coating a nickel alloy wire or ribbon with a uniform layer of barium and strontium carbonate paste. This then decomposes during the evacuation of the containing tube and further heating to above 1000 K causes oxygen vacancies and free barium atoms to be formed at the metal–oxide interface, making the material into a strong n-type semiconductor (see Section 14.1). The Fermi level is located near the bottom of the conduction band and gives a work function of about 1 eV. This, combined with surface states (see Section 14.4) which tilt the energy bands near the surface, makes the oxides as efficient emitters at 1000 K as tungsten is at 2000 K.

In these, as in all semiconductors, the work function is strongly temperature dependent and equation (10.2) does not therefore apply.

11.7 Schottky effect

When electrons are emitted from a heated metal, they will build up a space charge which tends to limit the emission. If an electric field $\mathscr{E}$ is applied to remove the electrons as they are emitted, then this has the effect of lowering the barrier as shown in Figure 11.6. The shape of the potential barrier outside the

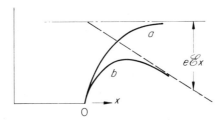

Figure 11.6. Lowering of potential barrier at metal surface due to electric field

surface (curve *a*) is given by $V(x) = -e^2/16\pi\varepsilon_0 x$ where x is the distance from the surface. At x, the barrier (curve *b*) is lowered by an amount $e\mathscr{E}x$, so that the potential curve will have a maximum at some distance from the surface. The barrier is thereby lowered so that the effective work function is reduced. This lowering of the barrier by an applied field is known as the *Schottky effect*.

If a very high strength electric field is applied, then the work function is decreased still more and the barrier is made so thin that quantum-mechanical tunnelling (Section 3.1) occurs. The current densities are then very large and relatively independent of temperature, giving *field emission*.

11.8 Thermoelectricity

11.8.1 Seebeck potential—thermocouples

It was shown in Section 10.6, that, because there is a variation in the density of electron states $N(E)$ with energy E, the value E_0 of the Fermi level will vary with temperature, being greater for higher temperatures, except near the top of a conducting band (which will not apply for most metals).

Hence, if a rod of conducting material has one end maintained at a higher temperature than the other, the electrons will at the hot end have, on average, higher kinetic energies. (The difference between the energy of an electron and the energy level of the bottom of the conducting band is the kinetic energy.) Hence electrons would tend to drift to the cold end, thereby making it negatively charged relative to the hot end. This induced voltage gradient will cause a current to flow in the reverse direction, equilibrium being reached when the current is equal and opposite to the drift of electrons due to the temperature difference. The induced voltage will then reach a constant value for any one temperature.

At that stage, the average energy of electrons along the rod will be constant, i.e. the lower kinetic energy at the cold end is balanced by a higher potential energy due to the negative potential. The Fermi levels at the ends have the same energy values and the energy bands have moved relatively due to the potential difference.

The voltage difference can be measured only by connecting the ends of the rod to a volt meter and the connecting wires will have a voltage gradient for the same reason. In general, the voltages induced in two different metals due to the same temperature difference are not of the same value so that a net voltage can be measured.

This thermally induced voltage V_{12}, the *Seebeck* potential, is used for determining temperatures by thermocouples. In usual applications, one junction is at a known temperature (usually 0 °C or the ambient temperature) and the

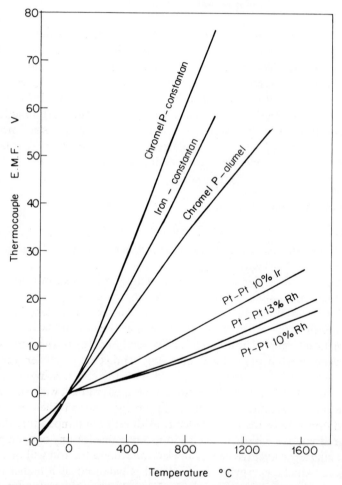

Figure 11.7. Thermocouple E.M.F.'s for some common thermocouple pairs. Chromel P = 90 Ni 10 Cr, Alumel = 95 Ni 5(Al + Si + Mn), Constantan = 55 Ni 45 Cu

Conducting Materials

other is at the temperature to be measured. The values of the Seebeck potential for a number of the more commonly used thermocouple materials are shown in Figure 11.7. These are approximately linear curves and single valued for the temperature ranges shown, but this is not necessarily the case for all pairs of metals. A copper–iron couple has a temperature–voltage relationship as shown in Figure 11.8.

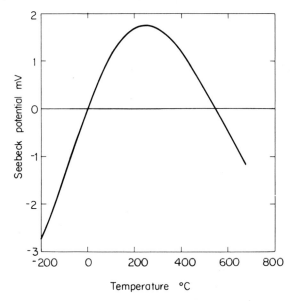

Figure 11.8. Seebeck potential versus temperature for copper–iron thermocouple

Because the voltage difference depends upon the variation of Fermi level with temperature in each material, then the Seebeck potential for one pair of metals is related to the Seebeck potential of each of them with a third metal by

$$V_{12} = V_{13} - V_{23}$$

11.8.2 Thomson effect

If a current flows along a rod which has a temperature gradient between its ends, then the energies of the electrons vary during their passage along the rod, which results in evolution or absorption of heat—the *Thompson effect*. If the electrical current is from the hot to the cold end, the electrons flow from a cold to a hot region and increase their energy by absorbing heat. Current flow in the reverse direction causes evolution of heat.

If dQ/dt is the rate of evolution of heat per unit volume, J is the current density and dT/dx is the thermal gradient, then

$$\frac{dQ}{dt} = -\mu_T J \frac{dT}{dx}$$

where μ_T is the *Thomson coefficient*.

Any Thomson heating or cooling effect is in addition to ohmic resistance heating.

11.8.3 Peltier effect

Heat is also absorbed or evolved when a current is passed through a thermocouple junction, the absorption or evolution depending upon the direction of the current. If dQ/dt is the rate at which heat is evolved when a current I flows from metal 1 to metal 2

$$\frac{dQ}{dt} = \pi_{12} I$$

where π_{12} is the *Peltier coefficient*.

Obviously

$$\pi_{12} = -\pi_{21}$$

This effect can be used as a heat pump by cooling one junction at the expense of heating the other using electric power from an external source. For metals the overall thermodynamic efficiency would be negligible, but by using semiconducting materials which have much higher Peltier coefficients it is possible to construct a thermoelectric refrigerator.

11.8.4 Kelvin relations

The three quantities just discussed are not independent and if one is known the others can be calculated. If we define the *thermoelectric power* of a metal as

$$S = \frac{dE_0}{dT}$$

then the variation of Seeback potential with temperature is given by

$$\frac{dV_{12}}{dT} = S_{12} = S_1 - S_2$$

The Kelvin relations for the Thomson and Peltier coefficients in terms of the thermoelectric power are

$$\mu_T = T \frac{dS}{dT}$$

and

$$\pi_{12} = TS_{12}$$

Questions for Chapter 11

1. What is the effect upon the energy levels when isolated atoms are brought together to form a crystal? Show by diagrams the distribution of the levels in crystals of (a) sodium, (b) magnesium, (c) diamond and (d) silicon and the extent to which they are occupied at room temperature. Comment upon the effect of these distributions on electrical conductivity. [MST]

2. A uniform silver wire has a resistivity of $1 \cdot 54 \times 10^{-8}$ Ωm at room temperature. For a potential gradient along the wire of 1 V m^{-1}, calculate the mobility and the average drift velocity of the electrons, assuming there are $5 \cdot 8 \times 10^{28}$ conduction electrons m^{-3}.

   Briefly explain why the current density in a metal is constant for a given potential gradient (Ohm's law) although the conduction electrons can be regarded as a perfect gas. [P].

3. Discuss the theory of metallic conductivity with particular reference to (a) Ohm's law, (b) the Wiedemann–Franz Law and (c) Matthiessen's rule.

   At room temperature the electrical resistivities of copper and of a complex iron–chromium alloy are $1 \cdot 7 \times 10^{-8}$ and $1 \cdot 4 \times 10^{-6}$ ohm m, respectively. Assuming that the Wiedemann–Franz law holds for both materials, find the electronic contribution to thermal conductivity in each case. Suggest reasons for any difference in the values obtained. [MST]

4. Discuss briefly the mechanism of thermionic emission and show, qualitatively, how the work function of a surface is affected by an external field.

   In a test on a tungsten filament, the following data were obtained:

Temperature (K)	2700	2500	2300	2200
Saturated emission (A m^{-2})$10^4 \times$	1·0	0·3	0·04	0·012

   (a) Determine the work function and the constant A in the Richardson–Dushman equation $I = AT^2 e^{-e\phi/kT}$.

   (b) By how many electron-volts must the work function change in order to reduce the emission at 2500 K by 10 per cent? What factors might cause such a change in the work function? [MST]

5. Determine the reduction in work function and the position of the potential barrier with respect to a cathode surface when exposed to a field of 10^6 V m^{-1}.

6. Distinguish between *field-emission* and *field-enhanced thermionic emission*.

A long fine tungsten wire ends in a hemisphere of radius 10^{-6} m. A hemispherical collector electrode is situated at a distance of 0·1 m from the tungsten hemisphere, so that the two are concentric and similarly placed. The collector is at a positive potential of 10 kV with respect to the tungsten.

Estimate the reduction in work function of the end of the tungsten wire. Discuss qualitatively the order of potential difference required to give field emission.

(The work function of tungsten = 4·45 eV = $7·1 \times 10^{-19}$ J. The lattice spacing for tungsten is of the order of 5×10^{-10} m.) [EST]

CHAPTER TWELVE

Insulating Materials

12.1 Dielectrics

In materials which are electrical insulators, often referred to as dielectrics, all the electrons are bound so tightly to their respective atomic nuclei that electrical conduction by electrons cannot occur. In terms of the discussion in Chapter 10, the energy gaps are so wide that electrons cannot receive enough thermal energy at room temperature to be excited across the gaps in sufficient numbers to provide a measurable conductivity.

When an electric field is applied to a sample of such a material, then the positive charges will undergo small displacements in the direction of the field and the negative charges in the reverse direction. Because the charges are not free, the displacements are limited, but the net effect is equivalent to a series of electric dipoles oriented in the direction of the field. Electric charges then appear on the surfaces as illustrated in Figure 12.1.

Figure 12.1. Charges on surface of dielectric due to an applied electric field

This process is known as polarization, and can be due to various effects which are additive when more than one exist together in the same material.

In ionic crystals, an applied electric field causes the positive ions to be displaced in the direction of the field and the negative ions to be displaced in the reverse direction, giving *ionic polarization*.

In any atom, the electrons are in motion around the nucleus, but due to an applied field, their centre of motion will be displaced relative to the nucleus, each atom thus becoming a dipole, giving *electronic polarization*.

The thermal motion of atoms and molecules has little effect upon the orientation of these *induced dipoles* and consequently the degree of ionic and electronic polarization is nearly independent of temperature.

In some materials, complex ions or molecules already possess dipole moments (see Sections 5.5 and 5.6) which, in the absence of a field, are randomly oriented so that the material has zero net moment. An external field tends to orient these dipoles, giving *dipolar polarization*, but the thermal agitation of the molecules tends to randomize the orientation. Thus this type of polarization, which only occurs in *polar materials*, is strongly temperature dependent.

A fourth effect, which occurs in heterogeneous materials containing two or more phases, is *interfacial polarization*. Due to differing conductivities of the different phases, application of an external electric field causes accumulations of charges at the interfaces between phases.

12.2 Dielectric constant

The presence of the surface charges due to polarization causes a reverse field inside the dielectric so that the original field is reduced. Suppose an electric field $\mathscr{E}_0$ perpendicular to two faces of a slab of dielectric, as shown in Figure 12.1, causes surface charges of magnitude P per unit area to appear on these faces. The field strength $\mathscr{E}$ inside the dielectric is then

$$\mathscr{E} = \mathscr{E}_0 - P/\varepsilon_0$$

where ε_0 is the permittivity of a vacuum and equals 8.854×10^{-12} F m^{-1}.

Consider a capacitor consisting of two parallel plates, each of area A, separated by a distance d in vacuo, which is connected to a voltage source of strength V (Figure 12.2(a)). Current flows from the voltage source until there is a

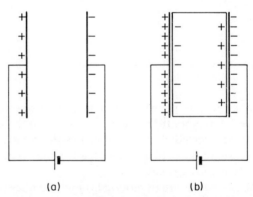

Figure 12.2. Capacitor **(a)** without, and **(b)** with dielectric between plates

Insulating Materials

charge of magnitude Q on each plate, so that

$$Q = CV \tag{12.1}$$

where C is the capacitance of the capacitor and is given by

$$C = \varepsilon_0 A/d \tag{12.2}$$

When a dielectric material is placed in the gap, it becomes polarized due to the field and charges appear on the surface with signs opposite to those on the adjacent plates (Figure 12.2(b)).

The voltage difference of the plates must remain constant because they are still connected to the source so that current flows from the source to build up an extra charge on each plate, this extra charge being equal to the final induced charge on the dielectric surface and of opposite sign. The total charge is then

$$Q' = C'V \tag{12.3}$$

where C' is the new capacitance of the capacitor and

$$C' = \varepsilon A/d \tag{12.4}$$

ε being the permittivity of the material.

It can be seen that

$$Q' = Q + PA \tag{12.5}$$

Hence by substituting equations (12.1) to (12.4) in equation (12.5)

$$\frac{\varepsilon A V}{d} = \frac{\varepsilon_0 A V}{d} + PA$$

or

$$P = (\varepsilon - \varepsilon_0)V/d$$
$$= (\varepsilon - \varepsilon_0)\mathscr{E}$$

where $\mathscr{E} \, (= V/d)$ is the electric field.

The ratio of the capacitances with and without the dielectric is known as the *relative permittivity* or *dielectric constant* ε_r, i.e.

$$\varepsilon_r = \frac{C'}{C} = \frac{\varepsilon}{\varepsilon_0}$$

and

$$P = \mathscr{E}(\varepsilon_r - 1)\varepsilon_0 \tag{12.6}$$

Also the ratio $P/(\varepsilon_0 \mathscr{E})$ is defined as the electrical susceptibility χ_e, so that

$$\chi_e = (\varepsilon_r - 1)$$

12.3 Polarizability

The forces which tend to cause polarization are those acting on the individual charges due to the local field. This local field will be the sum of the external field and that due to the surrounding dipoles in the medium. We calculate the internal field acting on an atom by considering it to be located at the centre of a spherical cavity in a dielectric inserted in a parallel-plate capacitor as in Figure 12.3(a).

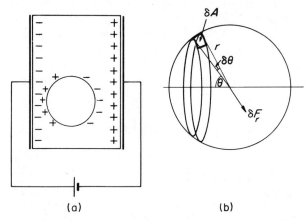

(a)　　　　　　　(b)

Figure 12.3. (a) Spherical cavity in dielectric used for calculation of internal local field, (b) element on surface of cavity

Charges appear on the surface of the cavity, the charge density at any point equalling $P \cos \theta$ where θ is the angle between the normal to the surface and the direction of the external field.

The charge on a surface element of area δA is $P \cos \theta \, \delta A$ and the force exerted by this charge on a unit positive charge at the centre of the cavity is

$$\delta F_r = \frac{P \cos \theta \, \delta A}{4\pi \varepsilon_0 r^2}$$

where r is the radius of the cavity (Figure 12.3(b)).

By symmetry, all components of forces such as δF_r would balance, except in the direction of the external field. The component of δF_r in that direction is

$$\delta F = \delta F_r \cos \theta$$
$$= \frac{P \cos^2 \theta \, \delta A}{4\pi \varepsilon_0 r^2}$$

Consider now that portion of the spherical surface which is cut off by cones of angles θ and $\theta + \delta\theta$ as in Figure 12.3(b). The area is

$$\delta A = 2\pi r^2 \sin \theta \, \delta \theta$$

Insulating Materials

so that the net force due to the charges on this area is

$$\delta F = \frac{P \cos^2 \theta}{4\pi\varepsilon_0 r^2} 2\pi r^2 \sin\theta\, \delta\theta \qquad (12.7)$$

It will be seen that when $\theta < 90°$, the charges are positive and when $\theta > 90°$, the charges are negative so that the force on a unit positive charge will be to the right for all values of θ.

The total force is obtained by integrating the expression in equation (12.7) over the whole sphere

$$F = \frac{P}{2\varepsilon_0} \int_0^\pi \cos^2\theta \sin\theta\, d\theta$$

$$= -\frac{P}{2\varepsilon_0}\left[\frac{\cos^3\theta}{3}\right]_0^\pi$$

$$= \frac{P}{3\varepsilon_0}$$

This, being the force that would be exerted on a unit positive charge, is also equal to the electric field due to the charges on the spherical surface.

We have also to consider the field at the centre of the cavity due to the dipoles which lie within the cavity. It can be shown that if the crystal symmetry of the material is cubic, then this field is zero. In crystals of lower symmetry, there is a field, but it is not large compared with $P/3\varepsilon_0$.

Hence, where cubic symmetry exists, the total field at a point within the dielectric is the sum of:

a: the field due to the charge density on the plates, i.e. $\mathscr{E} + P/\varepsilon_0$, where $\mathscr{E}$ would be the electric field if the dielectric were absent

b: the field due to the charge density on the dielectric surfaces facing the plates, i.e. $-P/\varepsilon_0$

c: the field at the centre of the spherical cavity as calculated above, i.e. $P/3\varepsilon_0$.

The local field $\mathscr{E}_{loc}$ is therefore given by

$$\mathscr{E}_{loc} = \mathscr{E} + P/3\varepsilon_0 \qquad (12.8)$$

The dipole moment p due to polarization of a single atom or a pair of ions or of a permanent dipole will be directly proportional to the electric field (at least for small fields) so that

$$p = \alpha \mathscr{E}_{loc}$$

where α is the electrical polarizability of the atom, ion pair or dipole.

The dipole moment of a unit cube of the dielectric will be the sum of the individual dipole moments within the unit volume and also will equal the surface charge density P. Hence

$$P = \sum n_j \alpha_j{}^j\mathscr{E}_{loc}$$

where there are n_j units of polarizability α_j for which the local field is $^j\mathscr{E}_{loc}$ and the summation is taken over all values of j.

For a dielectrically isotropic medium, $\mathscr{E}_{loc}$ is the same for all atoms and is given by equation (12.8), so that

$$P = \left(\mathscr{E} + \frac{P}{3\varepsilon_0}\right) \sum_j n_j \alpha_j \qquad (12.9)$$

By substituting equation (12.6) in equation (12.9), we get the *Clausius–Mosotti equation* relating dielectric constant to polarizabilities

$$\frac{\varepsilon_r - 1}{\varepsilon_r + 2} = \frac{1}{3\varepsilon_0} \sum_j n_j \alpha_j$$

In a diamond, for example, all atoms are alike, so that electronic polarizability α_e is the only factor to consider. Hence

$$\frac{\varepsilon_r - 1}{\varepsilon_r + 2} = \frac{1}{3\varepsilon_0} N \alpha_e$$

where there are N atoms per unit volume, and the value of α_e can be determined from an experimentally measured value of ε_r.

The dielectric constant equals the square of the refractive index n

$$\varepsilon_r = n^2$$

12.4 Temperature dependence

As stated in Section 12.1, thermal agitation has a strong influence upon the degree of polarization of permanent dipoles. In a fluid, the dipoles can rotate freely and so align themselves with an applied field. We will calculate the effect of thermal agitation upon the alignment.

If a dipole of moment μ lies at an angle θ to the field, then it makes a contribution $\mu \cos \theta$ to the polarization and the interaction energy with the field is $-\mu \mathscr{E} \cos \theta$. The solid angle contained between the cones of angle θ and $\theta + d\theta$ is $2\pi \sin \theta \, d\theta$, so that the probability of a dipole lying between these angles is proportional to

$$2\pi \sin \theta \, d\theta \, e^{-\mu \mathscr{E} \cos \theta / kT}$$

The average effective moment per dipole will be the sum of all the individual contributions to the polarization, divided by the total number of dipoles. That is

$$\bar{\mu} = \frac{\int_0^\pi \mu \cos \theta \, 2\pi \sin \theta \, e^{-\mu \mathscr{E} \cos \theta / kT} \, d\theta}{\int_0^\pi 2\pi \sin \theta \, e^{-\mu \mathscr{E} \cos \theta / kT} \, d\theta}$$

Insulating Materials

By substituting
$$a = \mu \mathscr{E}/kT$$
and
$$x = a \cos \theta$$
so that
$$dx = -a \sin \theta \, d\theta$$

$$\bar{\mu} = \frac{\mu \int_a^{-a} x e^{-x} dx}{a \int_a^{-a} e^{-x} dx}$$

whence
$$\frac{\bar{\mu}}{\mu} = \frac{e^a + e^{-a}}{e^a - e^{-a}} - \frac{1}{a}$$

which is called the *Langevin* function L(a), where
$$L(a) = \coth a - 1/a \qquad (12.10)$$

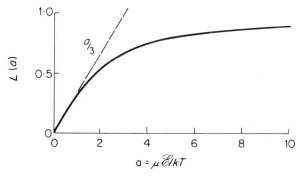

Figure 12.4. Langevin function

A graph of the variation of L(a) with a is shown in Figure 12.4. When a is small
$$L(a) \simeq a/3$$
so that the polarization is a linear function of the applied field. When $\mathscr{E}$ is very large, or T is very small, then $\bar{\mu}/\mu$ approaches the limiting value of one which corresponds to complete alignment of all the dipoles.

The breakdown voltage (see Section 12.7) of most materials is of the order of 2×10^7 V m^{-1} or less, and dipole moments are of the order of 2×10^{-29} C m or less, so that at room temperature and for practicable voltages, a will not exceed 0·1 at which value $L(a)$ still very closely equals $a/3$ and so

$$\bar{\mu} = \frac{\mu^2 \mathscr{E}}{3kT}$$

Hence, provided that the density does not change appreciably, the dielectric constant of a polar fluid will decrease with rising temperature as may be seen for water in Figure 12.5.

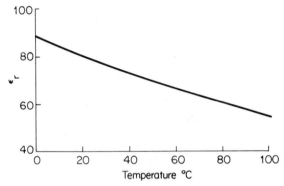

Figure 12.5. Variation of dielectric constant of water with temperature

In a solid, the dipoles are not completely free and the above is not strictly applicable. The polarization is, however, still temperature dependent and the function is of the form

$$\frac{\bar{\mu}}{\mu} = C \frac{\mu \mathscr{E}}{kT}$$

where C is a constant.

The total polarizability may then be written as

$$\alpha = \alpha_e + \alpha_i + C \frac{\mu^2}{kT}$$

where α_e and α_i are the electronic and ionic contributions, respectively.

The contribution of ionic effects, when present, is larger than that due to electrons, but dipolar polarization is of a still higher order of magnitude. The contribution of dipolar polarization to the total may be deduced from the variation of α with temperature.

12.5 Frequency response of polarization

If the polarizability is such that the restoring force is proportional to the displacement, then the behaviour in an alternating field is similar to that of a harmonic oscillator.

Insulating Materials

Due to the vibration of the electric charges, there will be energy emission due to radiation which is proportional to the charge velocity. Hence if x is the displacement of the charge, the differential equation for its motion under the influence of an alternating electric field of magnitude $\mathscr{E}_0$ and angular frequency ω will be

$$\frac{d^2x}{dt^2} + \gamma \frac{dx}{dt} + \omega_0^2 x = \frac{e}{m} \mathscr{E}_0 e^{i\omega t}$$

where ω_0 is the natural angular frequency of vibration of the charge and γ is the damping constant. This is the usual equation for a forced vibration and the solution is

$$x(t) = \frac{e}{m} \frac{\mathscr{E}_0 e^{i\omega t}}{\omega_0^2 - \omega^2 + i\gamma\omega}$$

If there are N polarizable units per unit volume, the susceptibility is

$$\chi = Nex/\varepsilon_0 \mathscr{E}$$

$$= \frac{Ne^2}{m\varepsilon_0} \left(\frac{1}{\omega_0^2 - \omega^2 + i\gamma\omega} \right)$$

$$= \frac{Ne^2}{m\varepsilon_0} \left[\frac{\omega_0^2 - \omega^2}{(\omega_0^2 - \omega^2)^2 + \gamma^2\omega^2} - i \frac{\gamma\omega}{(\omega_0^2 - \omega^2)^2 + \gamma^2\omega^2} \right]$$

$$= \chi' - i\chi''$$

where χ' and χ'' are real and imaginary components of the susceptibility, or the components of polarization that are in-phase and out-of-phase, respectively, with the exciting field.

Also the dielectric constant can be considered to be a complex quantity ε with real and imaginary parts ε' and ε'', respectively

$$\varepsilon = \varepsilon' - i\varepsilon''$$

$$= \varepsilon_0(1 + \chi' - i\chi'')$$

Typical variations of χ' and χ'' with frequency are shown in Figure 12.6. In a static field, $\omega = 0$ and

$$\chi_s = \frac{Ne^2}{m\varepsilon_0 \omega_0^2}$$

At low frequencies χ' is approximately equal to χ_s and increases with increasing frequency, but at values of ω near ω_0 it is strongly frequency dependent, becoming zero at $\omega = \omega_0$, undergoing a change of sign for $\omega > \omega_0$ and eventually becoming small for high values of ω.

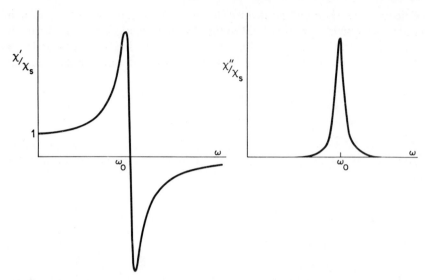

Figure 12.6. Typical variation of components of complex susceptibility with temperature. The ratio of each component to the static susceptibility is given

χ'' becomes significant when ω approaches ω_0 and is a maximum for $\omega = \omega_0$. The energy loss is a maximum at this frequency and is known as *resonance absorption*.

When electromagnetic radiation is passed through a dielectric material, the dipoles will experience a varying electric field of the frequency of the radiation.

If this frequency is lower than the resonant frequency of the dipoles, there will be little absorption and the material appears transparent. When the frequency of the radiation approaches the resonant frequency, there will be strong absorption.

For electronic polarization, the natural frequency $\omega_0/2\pi$ is approximately 10^{15} Hz which is the frequency of radiation in the visible or near-visible range (see Figure 2.2). Hence materials exhibiting only electronic polarization may have an absorption band in that region.

The natural frequency for ionic polarization is much lower, being of the order of 10^{13} Hz, which corresponds to infrared radiation.

When the radiation frequency is so much higher than that of the resonant frequency that the medium is no longer absorbing, the material will be totally reflecting for as long as $\chi < -1$ (i.e. ε_r is negative) and then be transparent again at the highest frequencies.

When dielectrics are subjected to alternating electric fields, the energy losses due to the damping of dipole moments and also due to any leakage currents (see Section 12.6) represent wasted energy. When used as a dielectric medium in a capacitor, the polarization is not exactly in phase with an applied alternating

Insulating Materials

voltage and the current will not be exactly 90° out-of-phase with the voltage. The difference between the actual phase angle and 90° is the *loss angle* δ. The power factor is

$$\tan \delta = \varepsilon''/\varepsilon'$$

which represents the ratio of the energy lost per cycle to the maximum energy stored in the capacitor. The electrical energy lost will appear as heat and raise the temperature of the dielectric.

Obviously, the power factor will vary with frequency and a value measured at one frequency may not be applicable at a greatly different frequency.

12.6 Conductivity of dielectrics

As already stated, if the energy gap between a filled energy band and a vacant conducting band is sufficiently wide, then a negligible number of electrons are available as conductors at ordinary temperatures. In ionic insulators, any observed conductivity is usually due to ions. In a perfect crystal, all normal sites would be occupied by ions and conductivity can occur only if these exchange places by diffusion. If, however, vacancies or interstitial ions exist then conduction can occur more easily, in the former case by an ion moving into a vacancy and thereby creating a different vacancy (equivalent to the diffusion of a vacancy in the opposite direction) or in the latter by the diffusion of the interstitial ion.

If q is the activation energy for a process (see Section 9.8), then the diffusion rate is given by

$$D = D_0 \, e^{-q/kT}$$

where D_0 is a constant for that crystal. The ionic conductivity σ_i is given by

$$\sigma_i = eN\mu_i$$

where there are N monovalent ions per unit volume, each with mobility μ_i.

Now by Einstein's relationship (see Section 13.8)

$$\mu_i = \frac{eD}{kT}$$

so that

$$\sigma_i = \frac{e^2 N}{kT} D_0 \, e^{-q/kT}$$

i.e. the conductivity increases with increasing temperature. From the slope of graphs of $\log \sigma_i$ as a function of $1/T$, the activation energy can be determined. At low temperatures the value of q is found to be that appropriate to vacancy diffusion (i.e. vacancies must already be present) while at high temperatures, the value is that appropriate to the creation of vacancy–interstitial pairs.

The vacancies in crystals such as sodium chloride are thought to be due to the presence of metallic ions of higher valency. Thus one divalent ion could be substituted for two monovalent Na$^+$ ions, but would occupy the position of only one.

Surface conductivity is also very important in insulators and is affected greatly by humidity which may cause a very thin water layer to form on the surface together with dust, etc.

12.7 Electrical breakdown

When the electric field strength applied to a dielectric exceeds a critical value, a relatively large electric current flows so that the insulating properties of the dielectric are lost. The electric field strength at which this occurs is known as the *dielectric strength*. It is not constant but varies with the thickness of the material, becoming greater at larger thicknesses. There are several mechanisms by which breakdown can occur in dielectrics.

If heat is generated by ionic currents at a rate greater than it can be dissipated, the temperature rises—increasing the conductivity—known as *thermal breakdown*.

Conducting paths may be formed with the aid of imperfections, such as dislocations, or crystal interfaces where diffusion is more rapid, and give *electrolytic breakdown*.

Dipoles which surround a stressed region produce local imperfection states in the forbidden energy gap so that the ionization potential of the electrons is decreased and *dipole breakdown* can occur.

Any conducting electrons present due to impurity atoms or thermal excitation gain energy due to the applied field, and this energy is generally lost due to electron–atom collisions. If any electrons gain enough energy to create an electron–hole pair by an electron–electron collision, then the extra electrons can also be accelerated and collide in a similar manner giving *collision breakdown*.

Also if any gas bubbles are present, the gas molecules are ionized by much lower electric fields ($\sim 10^6$ V m^{-1}) than those required for ionization in the solid ($\sim 10^8$ V m^{-1}). Any gas ions so formed are accelerated and when they strike the solid may produce electron–hole pairs giving *gas-discharge breakdown*.

Dielectric strength is reduced by moisture, elevated temperatures, ageing, etc.

12.8 Insulating materials

A material which is to be used primarily for insulating purposes should have a low dielectric constant to keep the capacitance between conductors low and a low power factor to reduce heating effects, but a high insulation resistance and a high dielectric strength.

When, however, dielectrics are used in capacitors, then a high dielectric constant is usually desirable so that the capacitance for a constant volume is large.

Non-polar liquids, such as certain oils, are used as liquid dielectrics.

Ceramics and high polymers are used as solid dielectrics. The ceramic materials include glass, quartz, mica, alumina (Al_2O_3) and magnesia (MgO) which have dielectric constants in the range 4–10.

Rutile (TiO_2) has a dielectric constant of a much higher order. It has a tetragonal crystal structure and anisotropic dielectric properties. The relative permittivity has values of 170 and 70 in directions parallel and perpendicular, respectively, to the c axis and a value of about 100 in polycrystalline material. It has a very low power factor even at radio frequencies and so can be used for capacitors at such frequencies and where small size is especially desirable. The temperature coefficient of variation of the permittivity of rutile has a large negative value of $7 \cdot 5 \times 10^{-4} \, K^{-1}$. The value of this temperature coefficient can be varied by small changes in chemical composition. Capacitors using rutile as the dielectric medium will therefore have a negative coefficient of variation of capacitance with temperature and can be used in circuits where other components have positive temperature coefficients to give some degree of temperature compensation.

Various titanates which have higher permittivities are considered in Section 12.15.

Polymers are generally insulators because the interatomic bonding is entirely covalent or Van der Waals secondary bonding. This, combined with the ease with which they can be moulded to shape, makes them extremely suitable for use in electrical equipment. The dielectric constant depends upon the extent to which the molecules can be polarized. Linear molecules are sufficiently flexible to orient themselves in an electric field, so that those which are polar can have high dielectric constants. As stated in Section 12.5, the dielectric constant varies with frequency, and for dipolar polarization, it falls off at high frequencies. Thus the dielectric constant of polyvinyl chloride (p.v.c.), which has unsymmetrical long chain molecules, falls from a high value at low frequencies to a much smaller value at ultrahigh frequencies. On the other hand, polymers with symmetrical molecules, such as polythene and polytetrafluoroethylene (p.t.f.e.), in which the polarization is entirely electronic, have smaller dielectric constants which are constant over a very wide frequency range (see Figure 12.7). The thermosetting resins which form a three-dimensional network of covalent bonds have intermediate values of dielectric constant, of the order of 4–5.

The dielectric loss factor, which is important at high frequencies, is much less in non-polar polymers than in polar ones. Thus at 1 MHz, the loss in polythene or p.t.f.e. is about 1 per cent. of that in p.v.c.

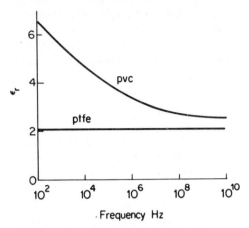

Figure 12.7. Variation of dielectric constant with frequency in a polar and a non-polar thermoplastic polymer

The fillers and plasticizers used in polymeric materials also affect the dielectric constant and dielectric loss. Cellulose fillers give an absorption peak at about 10^9 Hz, whereas mineral fillers do not. Many plasticizers are polar compounds which would increase the dielectric constant of a non-polar polymer. Even non-polar plasticizers increase the dielectric constant due to interfacial polarization.

Where good insulating properties are the sole electrical requirement, and flexibility is not needed, then the materials which can be used include hard rubbers, hard thermoplastic materials like polystyrene, thermosetting resins, glass and ceramics. The organic materials have an advantage over the inorganic by being less brittle, but they cannot be used at high temperatures. Again some, such as hard rubber and phenolformaldehyde resins, are affected by ultraviolet light and exposure.

For wire and cable insulation, flexibility is a necessity. Rubber was once used almost exclusively, but is now widely replaced by polythene and plasticized p.v.c. which have comparable dielectric strength and insulation resistance and are superior as regards durability. Also, being thermoplastic materials, they are easily moulded as cable coverings.

For a.c. applications, low dielectric constant and low loss factor are desirable, and generally organic materials are superior to inorganic in these respects. At ultrahigh frequencies, the dielectric loss is so significant that only very low loss materials, such as pure polyethylene and p.t.f.e., can be used.

12.9 Piezoelectricity

When mechanical stress is applied to a crystal, it distorts, the manner of distortion depending upon the degree of symmetry. Molecules or ion groups which

Insulating Materials

are aligned parallel to each other are likely to undergo a small rotation as the crystal is strained. If the structure is *centrosymmetric*, i.e. there are centres of symmetry in the atomic pattern, then any displacement will not upset the symmetry and the overall charge distribution is not appreciably affected. This is demonstrated in Figure 12.8, where each molecule is shown as consisting of a distribution of charge symmetrical about the centre.

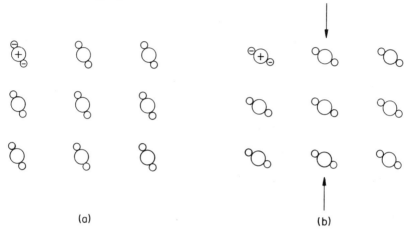

Figure 12.8. Displacement of centrosymmetric structure due to mechanical stress producing no polarization: (**a**) unstrained, (**b**) strained

Where, however, the structure is acentric, so that dipoles are present, then stressing will produce an asymmetric displacement which results in polarization, as shown in Figure 12.9. This change of polarization due to stressing is known as the *piezoelectric effect*.

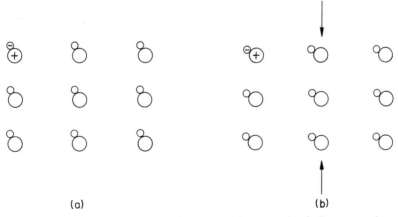

Figure 12.9. Displacement of acentric structure due to mechanical stress producing change in polarization: (**a**) unstrained, (**b**) strained

Conversely, when an electric field is applied to such a crystal, the dipoles will tend to line up with the field and so produce a mechanical strain. This is the *inverse piezoelectric effect*.

Crystals in which oriented dipoles are present are always piezoelectric. Those in which dipoles are so arranged that the moments cancel each other are piezoelectric only if the structure is not centrosymmetric.

A piezoelectric crystal is a mechanical–electrical *transducer* which has a linear response provided both the field and the strain are small.

The motion of the dipoles is subject to a damping term so that when an alternating electric field is applied to a piezoelectric crystal, the dipole displacements are not in phase with the field but usually lag. The phase difference varies with frequency and each crystal has a resonant frequency at which the energy absorption is a maximum.

This resonance is very sharp, so that such crystals can be used as resonators or as frequency controllers in oscillators.

Rochelle salt (sodium potassium tartrate tetrahydrate $KNaC_4H_4O_6 \cdot 4H_2O$) has the strongest piezoelectric effect known. Quartz has a less strong effect but is one of the most ideal materials for practical use because of its excellent elastic qualities combined with great mechanical strength and durability. The resonant or natural frequency of a quartz crystal depends upon the size and the orientation relative to the crystallographic axes. By a suitable choice of dimensions when cutting a crystal a specimen can be prepared with any desired natural frequency in the range from 50 Hz to higher than 10^8 Hz and which is independent of temperature.

12.10 Electrostriction

The application of an electric field to a crystal causes small displacements of the positive and negative charges which give induced dipole moments. The field then interacts with these dipole moments to produce a distortion which is proportional to the square of the field strength. Hence the distortion is always in the same sense whatever the sign of the field, and the inverse effect does not exist. This quadratic relationship between field and deformation is known as *electrostriction*. It is a common property of all material whether in the gaseous, liquid or solid state and is present in piezoelectric materials although the effect is too small to be significant compared with the piezoelectric strains except for very high electric fields.

12.11 Pyroelectricity

If, when a crystal is heated, the interatomic distances increase in an asymmetric way then, in a manner similar to the piezoelectric effect, polarization

Insulating Materials

may result. Such a material is termed *pyroelectric* and the phenomenon is known as the *pyroelectric effect*. All pyroelectric materials are piezoelectric, but the converse does not necessarily apply.

Zinc sulphide is a compound which exists in two allotropic forms (see Figures 7.20(c) and (d)). The cubic form (zincblende) is stable at room temperature, while the hexagonal form (wurtzite) is stable above 1020 °C and metastable below. Both forms are piezoelectric. The thermal expansion of zincblende is isotropic, but that of wurtzite is not because of the lower degree of symmetry. Hence of the two, only wurtzite is pyroelectric.

12.12 Ferroelectricity

This phenomenon was first discovered during investigations of the piezoelectric properties of Rochelle salt. This material possesses a dielectric hysteresis effect, to some extent analogous to ferromagnetism (see Chapter 15), for which reason the name *ferroelectricity* was given, and not for any direct connection with iron.

The existence of the dielectric hysteresis implies *spontaneous polarization*, i.e. a polarization which persists when tne applied electric field is zero and which requires a certain reverse field to change its direction.

To possess this property, a crystal must have two alternative structures which are identical except for the orientation of the dipoles (Figure 12.10). As

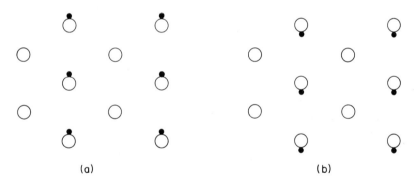

Figure 12.10. Alternative arrangements of ions in ferroelectric crystal. ○—anion, ●—cation

freshly prepared, ferroelectric crystals usually consist of multiple twins each of which is spontaneously polarized in a specific direction, but the direction of polarization of neighbouring twins is not parallel. That is *domains* exist and the net polarization is zero. Because these crystals are birefringent to light, the domain structure can be seen in polarized light.

When an electric field is applied to a crystal in a certain direction, the domains which have a direction of polarization more nearly parallel to the field will have a lower energy than antiparallel domains. Hence growth of the parallel domains at the expense of the others will result in a decrease of energy and so the domain boundaries move. The polarization will then no longer have an average value of zero. As $\mathscr{E}$ increases, the total polarization increases more rapidly (curve OA in Figure 12.11) until nearly all the domains are parallel.

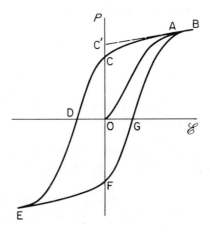

Figure 12.11. Hysteresis curve for ferroelectric material

Then the rate of increase decreases until saturation is reached when the crystal is a single domain (curve AB). The polarization is usually accompanied by a distortion in the form of elongation along the polarization direction. When the field is decreased to zero, the uncompensated charges on the surface of the crystal produce a reverse field which has a depolarizing effect and makes the uniform polarization unstable. A small change of domain boundary will, however, give enough decrease of net polarization to achieve stability, the remnant polarization being represented by C on Figure 12.11. The linear portion AB of the initial polarization curve, extrapolated backward to $\mathscr{E} = 0$ would cut the polarization axis at C', OC' then being the polarization of a single domain in a zero external field.

To reverse enough domains to make P again equal to zero, a reverse field OD, the *coercive field* must be applied. Further increase of the reverse field will give further shift of domain boundaries until saturation in the reverse direction is reached at E. Again reversing the field direction will give the curve EFGB, completing the hysteresis loop.

Insulating Materials

The exact shape of the experimentally determined hysteresis loop depends upon various factors which include the specimen dimensions, the temperature, the perfection of the crystal structure and its thermal and electrical history. The curve shown in Figure 12.11 is now believed to represent not a true single crystal, but a superposition of the loops of smaller crystallites. Very good quality single crystals have hysteresis loops which are almost parallelograms with vertical sides (Figure 12.12), i.e. the change of polarization occurs suddenly at the coercive field.

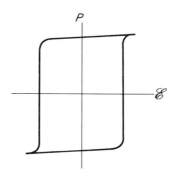

Figure 12.12. Hysteresis curve for single crystal

12.13 Effect of temperature

As the temperature is increased, the hysteresis loop changes its shape, the height decreasing slightly and the width decreasing considerably until at a temperature known as the *Curie temperature*, which is specific for each material, the loop has merged into a single line and the coercive field is zero. Thermal energy has assisted the applied field to overcome the potential barrier to reversal of polarization and at the Curie temperature is sufficient to 'randomize' the polarization directions unaided. Alternatively the structure may adopt a higher degree of symmetry above the Curie temperature.

Hence ferroelectric properties are absent at any temperature T above the Curie temperature T_c. The susceptibility is given by

$$\chi_e = \frac{C}{T - T_c}$$

where C is a constant, i.e. a similar relationship to that for ferromagnetic materials (see Section 15.6).

Most materials which are ferroelectric retain this property down to very low temperatures, but Rochelle salt is exceptional in not showing spontaneous polarization below $-20\,°C$.

12.14 Ferroelectric materials

Most of the known ferroelectric materials fall into three main groups, each with a specific type of crystal structure:

a. Rochelle salt and salts with a slight replacement of Na^+ or K^+ by other ions are ferroelectric. Most of the isomorphous tartrates (e.g. sodium ammonium tartrate) do not show this property above 0 °C. The dielectric constant of Rochelle salt reaches 4000 at the Curie temperature which is 24 °C. Rochelle salt has only one crystallographic direction of polarization.
b. Potassium dihydrogen phosphate (KH_2PO_4) and isomorphous dihydrogen phosphates and arsenates.
c. Barium titanate and many other compounds of the type ABO_3 (A and B representing metallic ions) with similar structure.

Of these, only barium titanate will be considered in any detail here.

12.15 Barium titanate

Above 120 °C, this substance, more correctly called barium titanium oxide because discrete titanate ions are not present, has a cubic cell structure as shown in Figure 12.13. It can be considered as a face-centred cubic structure with a

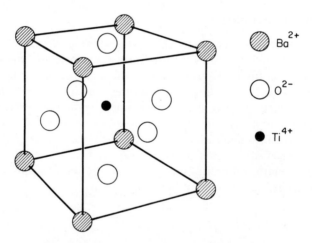

Figure 12.13. Unit cell of cubic barium titanate

close packing of the larger barium and oxygen ions, the smaller titanium ions occupying the octahedral interstices of the oxygen ions. Each barium ion has twelve equidistant oxygen ions and each titanium ion has six equidistant oxygen ions.

Insulating Materials

When cooled below the Curie point (120 °C), the structure changes to tetragonal with an axial ratio of about $c/a = 1.01$ and there are slight relative displacements of the ions as shown in Figure 12.14. The oxygen octahedra are

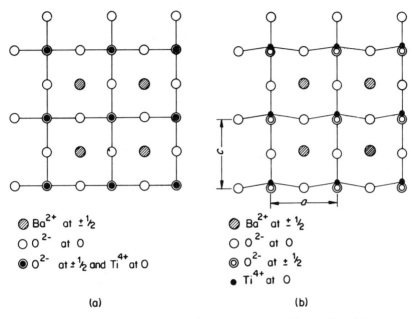

Figure 12.14. Projection of (a) cubic form, (b) tetragonal form of $BaTiO_3$ on to (010) plane. Heights of atoms above base plane are given in terms of unit cell dimensions. (Acknowledgments to Methuen)

themselves very little distorted, but are displaced about 0.08 Å in the c direction relative to the barium ions. The titanium ions are displaced about 0.05 Å in the reverse direction. Thus it will be seen that the structure has spontaneous polarization.

There are two further changes at lower temperatures, the structure becoming orthorhombic below 0 °C and rhombohedral below −90 °C. Both of these are spontaneously polarized, but whereas in the tetragonal structure the polarization was along one of the [100] directions of the corresponding cubic structure, in the lower temperature forms it is along a [110] and a [111] direction, respectively.

Barium titanate has a permittivity of 1400 at room temperature which falls slightly as the temperature rises until about 70 °C after which it rises to a value of 6000–7000 at 120 °C and above that temperature it falls rapidly. At any one temperature below 120 °C the permittivity varies as the applied field is varied, falling to half its initial value for a field strength of 4 MV m^{-1}. Although the dielectric constant is so high, the power factor is poor, so that in this respect it is inferior to rutile (see Section 12.8). When polarized, the material is piezoelectric.

Various chemical additions alter the properties. In particular, alloying with strontium titanate is used to reduce the variation of ε_r with temperature and with applied field.

Barium titanate is used in high ε capacitors in applications where a poor power factor can be tolerated, i.e. for low-frequency applications where space saving is important. It is also used in piezoelectric devices, which are polarized by raising the temperature above the transition point and applying a strong unidirectional field during cooling. These devices can be used as mechanical–electrical transducers in gramophone pick-ups and in strain gauges and also as electrical–mechanical transducers whereby electrical stimulation at the natural frequency of the specimen causes resonant vibrations.

Calcium titanate ($CaTiO_3$) and strontium titanate ($SrTiO_3$) are not ferroelectric above room temperature and have permittivities of about 150 and 240, respectively.

Questions for Chapter 12

1. What are the factors which can contribute to polarization in dielectric materials? Define *dielectric constant*.

    What are the effects of (*a*) temperature, (*b*) frequency of applied field upon the dielectric constant of materials? [MST]

2. Solid argon has a face-centred cubic crystal lattice with unit cell dimension 5·42 Å. The electronic polarization is $1·43 \times 10^{-40}$ F m^{-2}. Calculate the susceptibility and the relative permittivity.

3. By taking values from Figure 12.5, estimate the dipole moment of a water molecule.

4. For a dielectric with the susceptibility–frequency relationship given by Figure 12.6, sketch the variation with frequency of (*a*) the loss angle δ and (*b*) the fraction of incident electromagnetic radiation which will be transmitted.

CHAPTER THIRTEEN

Semiconductors

13.1 The energy bands of diamond

The spread of the energy bands of the valence electrons of atoms as they are brought together to form a crystal has been considered in Chapter 10. The discussion there centred on metals in which the valence electrons were virtually free to move anywhere within the crystal, but non-metals can be considered in a similar way.

Carbon, with the electron configuration $(1s)^2(2s)^2(2p)^2$, has four valence electrons per atom. When the carbon atoms are brought together to form diamond, these valence electrons form four covalent bonds to each atom directed in a tetrahedral manner and building up a structure as shown in Figures 7.18 and 7.19. Diamond is a very stable structure, that is the energies of the electrons in the covalent bonds are very low compared with their energies in the free atoms.

The $2s$ and $2p$ levels of the isolated atom contain two and six quantum states, respectively, and as the interatomic distance is decreased, each energy level widens into a band as shown in Figure 13.1. The energy bands first overlap and

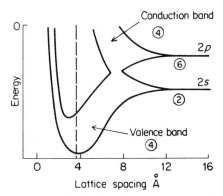

Figure 13.1. Energy band structure of diamond. Ringed numbers denote electrons per atom in each band

then at smaller interatomic distances split again into two bands, each of which contains four electron states per atom. The lower band corresponds to the energies of the electrons in the covalent bonds, which as shown in Section 5.6, consist of hybridized 2s and 2p orbitals. The upper band corresponds to electrons which can move freely in the crystal. The two bands are known respectively as the *valence band* and the *conduction band*. The reason for the latter name will become apparent in the next section.

There are just sufficient valence electrons to fill the lower band and this, being the state of lowest energy, is the stable configuration. At the standard interatomic distance the 5·3 eV energy gap between bands is so large that electrons cannot be taken across it by thermal excitation at ordinary temperatures. Hence, carbon in the form of diamond is an insulator.

13.2 Group IV semiconductors

Silicon and germanium atoms have outer electron structures similar to that of carbon with four electrons in the M and N shells, respectively. They also crystallize in the diamond form, i.e. with four covalent bonds per atom. The electron energy bands split in a similar manner as atoms are brought together, but the separation of the energy bands at the equilibrium distance is much less, being about 1·1 eV for silicon and 0·67 eV for germanium. At 0 K, all the electrons will occupy states in the valence band so that the crystals will be insulators, but at higher temperatures, some electrons can receive enough thermal energy to be excited into the conduction band. Then they have immediately adjacent vacant states and so can act as conductors of electricity. Hence the first empty band in such a structure is called a conduction band. The number of electrons available is obviously a function both of the temperature and of the narrowness of the energy gap, but is much less than would be present in a true metal, so that the conductivity is much lower. As stated in Section 10.10, materials of this type are distinguished from metals and true insulators by being called semiconductors.

Tin, with a similar valence electron arrangement, adopts the diamond structure in one allotropic form—gray tin—which is stable only below 18 °C, above that temperature adopting a body-centred tetragonal structure and having metallic properties. The energy gap of gray tin is much smaller than those of silicon and germanium, being 0·08 eV.

Other types of semiconducting materials are considered in Chapter 14.

When, due to the action of light or heat, some electrons are excited across the forbidden energy gap to occupy states in the conduction band, unoccupied states are left in the valence band. These are called *holes*. Since the crystal as a whole is electrically neutral, a hole is in effect a small positively charged region (see Figure 13.2).

Semiconductors

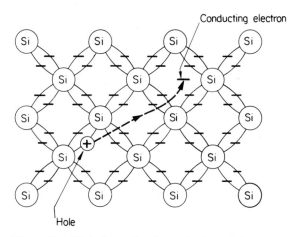

Figure 13.2. Hole formation by excitation of electron to conduction band

An electron may move into the vacant site from a nearby covalent bond, leaving a vacant site there. Hence 'holes' can move and contribute to the total conductivity.

When the random movements bring a conducting electron to a hole, it will jump back to the valence-bond position, with a release of the corresponding amount of energy. The rate of this recombination will be a direct function of the densities of holes and conducting electrons, whereas the rate of formation of holes and conducting electrons increases with increasing temperature, so that at any particular temperature, an equilibrium concentration of holes and conducting electrons will be established.

13.3 Density of states in a semiconductor

It was shown in Section 3.4 that the relationship between energy and the wave number away from the boundaries of the energy bands, i.e. where the Fermi surfaces are spheres, was given by

$$E = \frac{h^2 k^2}{8\pi^2 m} \tag{13.1}$$

and in equation (10.2) that the density of states is given by

$$N(E) = 4\pi \left(\frac{2m}{h^2}\right)^{3/2} E^{1/2} \tag{13.2}$$

When, however the value of k approaches a boundary of the Brillouin zone, the energy is no longer given by equation (13.1), but its value is strongly influenced by the periodic lattice. The shape of the E–k curve is a function of the crystal lattice and cannot, in general, be calculated but is determined indirectly by various experimental methods.

It can be shown that electrons which occupy states near the edges of the bands behave very similarly to free electrons and many of the equations already derived can be used directly by substituting for the mass m another quantity m^*, called the *effective mass* and defined by

$$m^* = \frac{h^2}{4\pi^2} \bigg/ \frac{d^2 E}{dk^2}$$

where $d^2 E/dk^2$ has the value determined from the curvature of the actual E–k curve near the discontinuities. Hence m^* is a function of the periodic nature of the lattice. Because the E–k curve may be different for different crystallographic directions, the value of m^* may vary with the direction of the wave vector $\mathbf{k}$. In considering the overall properties of materials, it is necessary to use a suitable average value of m^*, which can be determined experimentally.

At the top of the valence band, the E–k curve has curvature of opposite sign from that at the bottom of the conduction band so that m^* is negative. However, the energies of holes are of negative sign relative to the energies of electrons, so that the same equations can apply to holes in the valence band as to electrons in the conduction band.

In the simple band model, the energy of an electron in the conduction band is

$$E = E_c + \frac{h^2 k^2}{8\pi^2 m^*}$$

and the effective density of states, by analogy with equation (13.2), is

$$N(E) = 4\pi \left(\frac{2m^*}{h^2}\right)^{3/2} (E - E_c)^{1/2}$$

where E_c is the energy at the bottom of the conduction band.

The number of electrons dn per unit volume of semiconductor which lie in the energy range E to $E + dE$ is the product of the number of available states $N(E)\,dE$ and the probability that a state is occupied, i.e. the Fermi function, so that

$$dn = N(E) p(E) \, dE$$

The total number of electrons in unit volume in the conduction band is given by integrating this expression and is

$$n = \int_{E_c}^{\infty} N(E) p(E) \, dE$$

Semiconductors

In an intrinsic semiconductor, the Fermi level lies near the middle of the energy gap (see p. 200) and so $(E_c - E_0)$ is about 0·55 and 0·34 eV in silicon and germanium, respectively. Over the range of integration therefore, $(E - E_0)/kT$ will, at 300 K, be at least 14, so that $e^{(E-E_0)/kT} \gg 1$ and the Fermi function approximates to the Boltzmann distribution $e^{-(E-E_0)/kT}$.

The distribution of available states, the Fermi function for 600 K and the electron and hole densities in germanium are shown in Figure 13.3. It will be

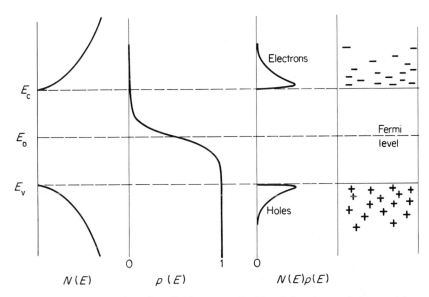

Figure 13.3. The density of available states, the Fermi function and the resulting electron and hole densities corresponding to 600 K in intrinsic germanium. The electron and hole densities are to a scale 500 times that of $N(E)$. The right-hand diagram shows the conventional manner of representing the electrons and holes in the conduction and valence bands

seen that the conducting electrons are confined to energy levels only just above and the holes to energy levels only just below the forbidden band. Because the spread of occupied states is small, except at very high temperatures, the available levels may be considered to be concentrated at the top and bottom of the forbidden band with effective densities N_c and N_v of available states at these energy levels. It can be shown that these effective densities are given by

$$N_c = 2\left(\frac{2\pi m_c^* kT}{h^2}\right)^{3/2}$$

and

$$N_v = 2\left(\frac{2\pi m_v^* kT}{h^2}\right)^{3/2}$$

where m_c^* and m_v^* are the effective masses of electrons and holes in the conduction and valence bands, respectively. Then

$$n = N_c e^{-(E_c - E_0)/kT} \qquad (13.3)$$

and

$$p = N_v e^{-(E_0 - E_v)/kT} \qquad (13.4)$$

where E_v is the energy of the top of the valence band.

In a pure semiconductor, like silicon or germanium, there are, at 0 K, enough electrons to fill exactly all the covalent bonds, or, in other words, to fill exactly the valence band. At higher temperatures, a hole is created for each electron raised to the conduction band, and hence

$$n = p$$

Then

$$\frac{N_v}{N_c} = e^{-(E_c + E_v - 2E_0)/kT}$$

By taking logarithms of both sides, this becomes

$$kT \log_e \frac{N_v}{N_c} = -(E_c + E_v) + 2E_0$$

or

$$E_0 = \frac{E_c + E_v}{2} + \frac{kT}{2} \log_e \frac{N_v}{N_c}$$

$$= \frac{E_c + E_v}{2} + \frac{kT}{2} \log_e \frac{m_v^*}{m_c^*}$$

If m_v^* and m_c^* are equal, then the Fermi level lies in the middle of the gap. If they are unequal, then the Fermi level is displaced from this value but only by a small amount because the second term is always much smaller than the first at ordinary temperatures.

13.4 Effect of impurity atoms in a semiconductor

A pure semiconductor as just described is known as an *intrinsic semiconductor*. Because the characteristics of a semiconducting material can be changed by the

Semiconductors

introduction of minute amounts of impurity (e.g. less than one part in a million) the degree of purification necessary to get intrinsic behaviour is extremely high and very specialized techniques are necessary to achieve this (see Section 14.2). When impurities are deliberately added to a pure semiconductor in controlled amounts for the purpose of modifying the properties, the material is said to be doped.

Consider first the doping of a quadrivalent semiconductor with a pentavalent element, e.g. antimony added to germanium. The antimony atoms will go into substitutional solid solution, i.e. they take positions that would otherwise be occupied by germanium atoms and do not change the crystal lattice. Four of the valence electrons of each antimony atom will go into the four covalent bonds with the neighbouring germanium atoms, but the fifth will be superfluous. The presence of the antimony atom also disturbs the general potential field in the crystal and acts as if an extra proton were introduced at this point into an otherwise regular periodic field. The field due to this extra charge is exactly the same as that for an isolated proton except that it is reduced in magnitude due to the dielectric constant of the semiconductor. The fifth electron will be attracted by the extra charge and move in its field (Figure 13.4). The energy levels that it can

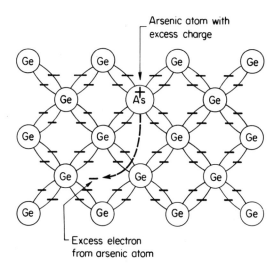

Figure 13.4. Schematic diagram of donor atom in germanium

occupy will be similar to those an electron can occupy in a hydrogen atom with due allowance made for the effect of the dielectric constant ε_r of the medium and for the effective mass of the electron. The lowest possible energy state will be

$$\left(\frac{m^*}{m}\right)\frac{1}{\varepsilon_r^2}E_1$$

where E_1 is the ionization potential of hydrogen (= 13·6 eV). For germanium, $\varepsilon_r = 16$, so that if we assume $m^* = m$, then this energy would be about 0·05 eV. This is the ionization energy, or the energy necessary to detach the electron from the attractive field of its parent atom and is very small compared with the energy gap. In the unexcited state, the fifth electron will occupy a level slightly below the bottom of the conduction band and will remain in the vicinity of the parent atom. The energy level variation in the region of the impurity atom can be represented as in Figure 13.5. All the energy bands are lowered due to the excess charge of the donor and the energy level of the electron is shown.

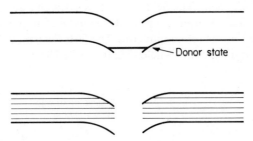

Figure 13.5. Effect of donor impurity atom on energy levels

The ionization energy of donor states in germanium has been found experimentally to be about 0·01 eV, so that one must assume $m^* \approx m/5$, a value in good agreement with other experimental measurements.

The corresponding Bohr orbit radius is

$$r = \varepsilon_r\left(\frac{m}{m^*}\right)0\cdot59\text{ Å}$$

i.e. about 47 Å for germanium. This is equal to several interatomic distances, so that the fifth electron is not closely associated with its parent atom. If the impurity atoms are sufficiently numerous so that the wave functions of the extra electrons overlap, then the energy levels will be modified, the ionization energy being reduced. Experimentally, this effect can be observed when the impurity density exceeds 10^{21} atoms m^{-3} while at densities greater than 10^{25} atoms m^{-3} the ionization potential is near zero and the semiconductor behaves like a metal, having reasonable conductivity at 0 K.

Semiconductors

The energy level of the unexcited extra electrons is known as a *donor level*. It has a specific energy E_d and also exists only in the vicinity of donor atoms, so that it is a *discrete localized state*. These states are usually shown as separate short lines on an energy diagram as in Figure 13.6(a). An electron occupying a donor level can be excited into the conduction band if it receives thermal or optical energy of 0·01 eV. At room temperature, most of the donor electrons are excited into the conduction band (Figure 13.6(b)).

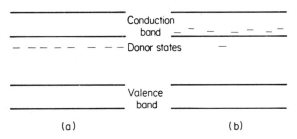

Figure 13.6. Representation of donor states and conduction electrons in *n*-type semiconductor on energy level diagram at (**a**) 0 K, (**b**) 300 K

For silicon, $m^* \approx 0.5\,m$ and $\varepsilon_r = 11.7$ so that the corresponding values are

Ionization energy 0·05 eV
Bohr radius   13·7 Å

Semiconductors with an excess of conducting electrons are known as *n*-type extrinsic semiconductors.

When a quadrivalent semiconductor is doped with a trivalent element, such as gallium added to germanium, one of the covalent bonds for each gallium atom is not satisfied, but can be completed by the transfer of an electron from a nearby valence bond, thus leaving a *hole* in the valence energy band (Figure 13.7). Because the added atom has accepted a valence electron from the crystal, such impurities are known as *acceptors*. Any semiconductor with excess holes is known as *p*-type.

The hole has an effective positive charge and the acceptor atom has an effective negative charge so that the motion of a hole near the impurity atom is analogous to that of an electron near a donor. The energy levels in the vicinity of the acceptor atom will be as shown in Figure 13.8, the acceptor state having energy E_a.

At 0 K, the hole will remain in the vicinity of the impurity atom, but at higher temperatures, electrons are excited from the valence band to fill the acceptor states (Figure 13.9). The holes left thereby in the valence band are free to move and give conductivity as described in Section 13.2.

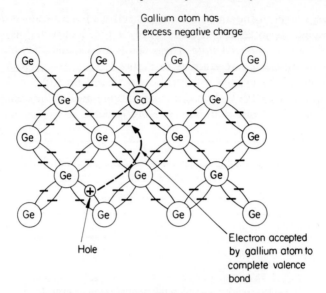

Figure 13.7. Schematic diagram of acceptor atom in germanium

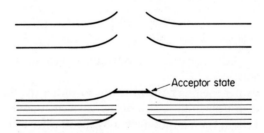

Figure 13.8. Effect of acceptor impurity atom on energy levels

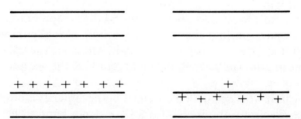

Figure 13.9. Acceptor states and conducting holes in *p*-type semiconductor at (**a**) 0 K, (**b**) 300 K

Semiconductors

If a semiconductor contains both donor and acceptor impurities, then the one in the majority determines whether there is n- or p-type behaviour.

13.5 Position of the Fermi level

In addition to thermal energy removing electrons from donor states into the conduction band or holes from acceptor states to the valence band, electrons will also be raised from the valence band to the conduction band as in an intrinsic conductor.

In an extrinsic semiconductor, the conducting electron and hole densities are given by equations (13.3) and (13.4), but the Fermi level is not necessarily near the middle of the forbidden energy band. These equations are, however, completely valid as long as $(E_c - E_0)$ and $(E_0 - E_v)$ are both considerably larger than kT. The product of n and p is then

$$np = N_c N_v \, e^{-(E_c - E_0)/kT} \, e^{-(E_0 - E_v)/kT}$$

$$= N_c N_v \, e^{-E_g/kT}$$

where E_g is the width of the forbidden energy gap (i.e. $E_c - E_v$). Thus np is a function of temperature only. For germanium at 300 K, for example

$$e^{-E_g/kT} = e^{-0.67/0.0258}$$

$$= e^{-25.9}$$

Also

$$N_c = N_v \approx 2.5 \times 10^{25} \, \text{m}^{-3}$$

so that for intrinsic material

$$n = p = \sqrt{np} \approx 2.5 \times 10^{25} \times e^{-12.95}$$

$$\approx 6 \times 10^{19} \, \text{m}^{-3}$$

For silicon the corresponding number is about $5 \times 10^{15} \, \text{m}^{-3}$. To show intrinsic behaviour at this temperature, the impurity concentration must be less than the value of n or p. As there are about 5×10^{28} atoms m^{-3}, the impurity concentration must be less than one atom in 10^{13} silicon atoms. For germanium the restriction is not so severe, the corresponding impurity concentration being one atom in 10^9 germanium atoms.

For a sample of n-type germanium containing 5×10^{21} donor atoms m^{-3}, the variation of the Fermi level with temperature is shown in Figure 13.10 and the numbers of conducting electrons and holes in Figure 13.11. Over the temperature range from 100 to 400 K the number of electrons is constant and equal to the number of donor atoms.

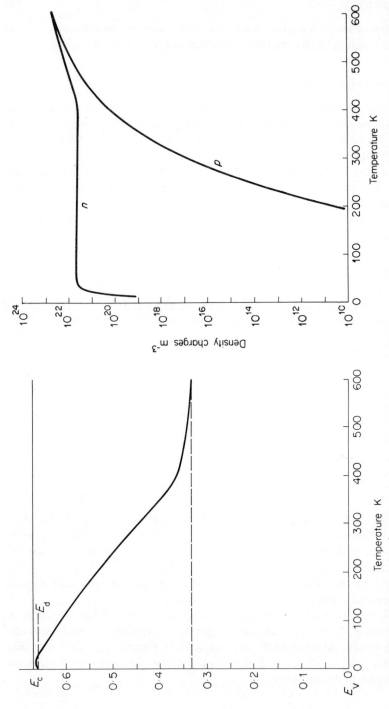

Figure 13.11. Variation with temperature of densities of conducting electrons and holes for *n*-type germanium doped as in Figure 13.10

Figure 13.10. Variation of Fermi level with temperature for *n*-type germanium doped with 5×10^{21} donor atoms m^{-3}.

Semiconductors

At very low temperatures, the electrons in donor states are excited into the conduction band and the Fermi level lies between E_d and E_c. Then for a wide range of temperature, n is constant at a value large compared with the intrinsic value so that p is small and can be neglected. At higher temperatures, the intrinsic value of n and p becomes comparable with the number of donors so that p is no longer negligible and the semiconductor shows intrinsic behaviour. As the number of electrons excited from the valence to the conduction band with rising temperature becomes significant relative to the number of donor atoms, the Fermi level falls and at a very high temperature approaches the value for an intrinsic semiconductor. Similarly, the Fermi level for a p-type semiconductor is mid-way between the acceptor level and the valence band at very low temperatures and rises with increasing temperature.

For very high doping concentrations the donor or acceptor level becomes closer to a limit of the energy band and at moderate temperatures, the Fermi level may lie within the conduction or valence band.

13.6 Conductivity in semiconductors

The factors which limit the mobility of electrons and hence the conductivity of a metal were discussed in Section 11.2. In a semiconductor, any holes present move in the same direction as the field and are positive charge contributions to the total current. Their contribution to the conductivity is

$$\sigma_p = ep\mu_p$$

where p is the number of holes per unit volume and μ_p is the mobility of a hole.

The total conductivity is then

$$\sigma_{\text{total}} = \sigma_n + \sigma_p$$
$$= en\mu_n + ep\mu_p$$

μ_p is generally smaller than μ_n.

We saw in Section 11.2 that in a metal the conductivity decreases as the temperature rises because of increased scattering due to thermal vibrations. In a semiconductor, the number of carriers increases with temperature so that the conductivity variation depends on more than one factor and may increase or decrease with temperature. The effect of the lattice vibrations is to give a variation of

$$\mu_L \propto 1/T^{3/2}$$

but the ionized impurities also cause a scattering which gives a mobility

$$\mu_I \propto T^{3/2}$$

These are independent processes and the total mobility is given by

$$\frac{1}{\mu} = \frac{1}{\mu_L} + \frac{1}{\mu_I}$$

$$= AT^{3/2} + BT^{-3/2}$$

Hence the mobility is determined at low temperatures by the impurity scattering and at high temperatures by the thermal vibration scattering. In purer specimens, the latter is the dominant factor over a wider temperature range. When, however, the numbers of carriers vary with temperature in an exponential manner, the conductivity is influenced by the exponential term far more than it is by either of the $\frac{3}{2}$ power laws. In an n-type semiconductor, for example, when the donor ionization energy is small, the donors are ionized at comparatively low temperatures and above this temperature n is almost constant until intrinsic conductivity appears. While n is constant, the variation of conductivity with temperature is determined by the mobility and follows the $T^{-3/2}$ or the $T^{3/2}$ law according to which is dominant at the temperature concerned.

Typical curves for the temperature variation of conductivity in a semiconducting material with various degrees of doping are shown in Figure 13.12.

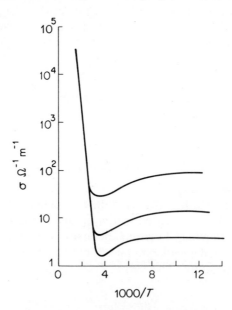

Figure 13.12. Typical temperature variation of conductivity in germanium with various degrees of doping. Lowest curve is for purest sample

Semiconductors

When intrinsic behaviour governs, the conductivity will be proportional to the number of carriers and since

$$n = p = N_c e^{-E_g/2kT}$$

a plot of log σ against $1/T$ should be a straight line of slope $-E_g/2k$, which will be common to all samples as shown on Figure 13.12. When the temperature is very low, the number of carriers will be governed by the ionization of donor electrons according to

$$n \propto e^{-(E_c - E_d)/kT}$$

and the line should have a slope $-(E_c - E_d)/2k$. One experimental method by which the energy gaps are determined is from the slopes of the conductivity curves.

In the intervening region, the conductivity will rise in the range in which the $T^{3/2}$ relationship dominates and at higher temperatures fall when the $T^{-3/2}$ relationship dominates until intrinsic conduction becomes the governing factor.

Because $\mu = e\tau/m^*$, then for a semiconductor with carriers of one sign only,

$$\sigma = \frac{ne^2\tau}{m^*}$$

Because the square of the charge appears in this equation, conductivity measurements will not reveal whether the current is due to electrons or holes. A further experiment which can determine this is described in the next section.

13.7 Hall effect

When a magnetic field is applied to a conductor which is carrying a current, the direction of the magnetic field not being parallel to the direction of the current, an electric field is set up in a direction which is perpendicular to both the current and the magnetic field. This is the *Hall effect* and it can be used to determine both the sign of the majority carriers and their density.

Consider a block of material of rectangular cross-section, width b and thickness t, as shown in Figure 13.13. Suppose that a current I flows along the length of the block. If n is the density of carriers, each with charge q, and if these have an average velocity $\bar{v}$ in the direction of the current, then

$$I = btnq\bar{v}$$

The current density will be

$$J = \frac{I}{bt} = nq\bar{v}$$

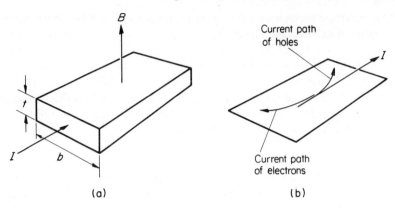

Figure 13.13. (a) Directions of fields and current in Hall effect, (b) electron and hole paths in relation to the electrical and mechanical fields

If a magnetic field with flux density B is applied in the thickness direction, then the charges will experience a force in the breadth direction which deflects them to one side of the conductor. The force is $Bq\bar{v}$ and will act in the same direction for both negative and positive charges, because if q changes sign, then so does $\bar{v}$. An electric charge will build up in the surface which will produce an electric field in the breadth direction. The build-up will continue until the force on the charges due to this field exactly balances that due to the magnetic field, so that the charges again travel directly along the length of the conductor. If this transverse electric field has the value $\mathscr{E}_H$

$$q\mathscr{E}_H = Bq\bar{v}$$
$$= \frac{BJ}{n}$$

or

$$\mathscr{E}_H = \frac{BJ}{nq}$$

The *Hall constant* of the material is defined as

$$R = \mathscr{E}_H/BJ$$

and in this case is

$$R = \frac{1}{nq} \qquad (13.5)$$

Semiconductors

For experimental purposes, a sample of semiconductor of the shape shown in Figure 13.14 can be used. It is of rectangular cross-section with enlarged ends which are used to connect the current leads, while the side projections are used for connections for potential measuring probes for both conductivity and Hall voltage measurements.

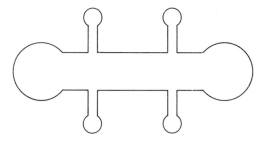

Figure 13.14. Possible shape of semiconductor sample for Hall conductivity measurements. The enlarged ends are for connecting the current leads and the side arms are for connections to measure voltage gradients along the specimen and Hall voltages

If the measured Hall voltage is V_H, then

$$\mathscr{E}_H = V_H/b$$

and

$$R = \frac{V_H t}{IB}$$

The sign of the carriers can be deduced from the sign of V_H and their density equals $1/Rq$.

In a semiconductor, the electron velocities have a Maxwell–Boltzmann distribution. The Hall constant, which is given by equation (13.5) for a metal, has to be modified to

$$R = \frac{3\pi}{8ne}$$

The conductivity is given by

$$\sigma = nq\mu$$

so that from a combination of conductivity and Hall effect measurements, the mobility can be determined.

To get the mobilities of both electrons and holes in a particular semiconducting material, it is necessary to use both n- and p-type specimens.

With intrinsic or near-intrinsic semiconductors both electrons and holes play a part and the relative mobilities also determine the Hall constant. It can be shown that

$$R = \frac{3\pi}{8} \frac{(nb^2 - p)}{(nb + p)^2}$$

where

$$b = \mu_n/\mu_p$$

is the ratio of the mobilities.

Hence the Hall constant may change sign with changing temperature as the carrier densities and the mobility ratio are temperature dependent.

13.8 Diffusion

If there is a non-uniform concentration of carriers in a semiconductor, then the carriers will diffuse in such a direction as to make the density more uniform. The diffusion constant D is defined by

$$J_D = -eD\frac{dn}{dx}$$

where J_D is the diffusion current density in the x direction. If at the same time there is an electric field which causes a current in the reverse direction, then

$$J_c = \sigma \mathscr{E}$$

If a steady-state condition is reached, then

$$ne\mu\mathscr{E} = -eD\frac{dn}{dx}$$

from which

$$n = C\,e^{-\mu\mathscr{E}x/D} \qquad (13.6)$$

Now due to the field $\mathscr{E}$, electrons which are a distance $\mathscr{E}x$ apart will have a difference in potential energy of $e\mathscr{E}x$, so that according to the Maxwell–Boltzmann statistics, the density of electrons should be given by

$$n = C\,e^{-e\mathscr{E}x/kT} \qquad (13.7)$$

By comparison of equations (13.6) and (13.7), the relationship between the diffusion constant and the mobility is found to be

$$D = \frac{kT}{e}\mu$$

which is known as the *Einstein* relation.

Semiconductors 213

Questions for Chapter 13

1. Define *Fermi level* and *Fermi temperature*.
   Show that, in an intrinsic semiconductor, the Fermi level lies half-way between the top of the valence band and the bottom of the conduction band. Assume that the Fermi function $f(E) = 1/[1 + e^{(E-E_f)/kT}]$, where E_f is the Fermi energy.
   It is proposed that a digital computer memory shall consist of a bar of germanium 0·02 m long with an applied field of 10^4 V m^{-1}. Electron pulses are injected into one end and reach a collector 0·015 m distant in 4·3 μs. Determine the mobility of the electrons and comment on the suitability of the device in terms of the number of pulses which can be stored in the bar in a given time. [MST]

2. What is meant by the terms *intrinsic, donor* and *acceptor* when applied to a semiconductor.
   What purity is necessary in a germanium crystal in order that there should be not more than ten impurity carriers for every intrinsic carrier? Assume that all impurity atoms are singly ionized and that the intrinsic carrier density is $2·5 \times 10^{19}$ carriers m^{-3}. Express your answer in parts per million by weight.
   (Atomic weight of Ge = 72·6; density = $5·36 \times 10^3$ kg m^{-3}. Atomic weight of impurity addition = 122.) [MST]

3. Explain briefly the differences between conductors, semiconductors and insulators.
   In order to produce a *p*-type semiconductor, silicon is to be doped with aluminium and the final alloy is to have a resistivity of 10 Ωm. Given that the mobility of holes in silicon is 0·35 m^2 V^{-1} s^{-1} and its density is $2·34 \times 10^3$ kg m^{-3}, calculate the amount of aluminium required in parts per million. It may be assumed that each impurity atom contributes one carrier. [MST]

4. Describe an experimental method which may be used to determine the sign and the density of charge carriers in a semiconductor in which conduction results almost entirely from carriers of one type.
   An *n*-type germanium sample is 2 mm wide and 0·2 mm thick. A current of 10 mA is passed longitudinally through the sample and a magnetic field of 0·1 Wb m^{-2} is directed perpendicular to the thickness. The magnitude of the Hall voltage developed is 1·0 mV. Calculate the magnitude of the Hall constant and the number of electrons per cubic metre. [MST]

5. Describe an experiment by which the mobility of electrons in a semiconductor may be determined.

Given that the mobility of conducting electrons in germanium is 0·38 $m^2 V^{-1} s^{-1}$ and its density is 5·36 × 10^3 kg m^{-3}, calculate the amount of arsenic, in parts per million by weight, with which the germanium must be doped to give an *n*-type semiconductor with a resistivity of 0·03 Ωm, if each arsenic atom loses an electron to the conducting band. [MST]

6. Explain why the electrical conductivity of a semiconductor usually increases while that of a metal decreases with increasing temperature.

Why does a 50/50 copper–nickel alloy have an electrical conductivity that is much smaller than that of either pure copper or pure nickel? [MST]

7. In a semiconductor containing *p* holes and *n* electrons per unit volume, the Hall constant is given, to a first approximation, by

$$R_H = \frac{p\mu_p^2 - n\mu_n^2}{e(p\mu_p + n\mu_n)^2}$$

where *e* is the electronic charge and μ_p and μ_n are, respectively, the mobilities of holes and electrons.

A certain intrinsic semiconductor has the following properties at room temperature, expressed in SI units:

$$R_H = -0.90, \quad \mu_p = 0.2, \quad \mu_n = 0.4$$

Estimate the resistivity of the material.

Estimate also the value that the resistivity would have had if the material had contained 2 × 10^{19} donor impurities per cubic metre and if all of these donors had been ionized.

It may be assumed that the classical approximation to the Fermi function is valid. [EST]

8. Sodium has a density of 971 kg m^{-3}, a resistivity of 4·7 × 10^{-8} Ωm and a Hall coefficient $-2.5 \times 10^{-10} m^3 C^{-1}$.

Calculate (*a*) the density of conducting electrons and (*b*) their mobility.

CHAPTER FOURTEEN

Semiconductor Materials and Devices

14.1 Semiconductor materials

14.1.1 *Elements*

The Group IV semiconductors, silicon and germanium, have been used in the preceding chapter as examples in describing the fundamentals of semiconductor behaviour. The other elements of that group with a similar crystal structure are carbon in the form of diamond and gray tin. Diamond has such a large energy gap that it is an insulator at room temperature, but becomes an intrinsic semiconductor at about 100 °C. Gray tin is stable only at low temperatures and has a very small energy gap.

The degree of purity required to give intrinsic behaviour at room temperature is less than one foreign atom in 10^{12} for silicon and 1 atom in 10^9 for germanium, so that germanium is often more convenient to use because of the lower degree of purification required. Also the carrier mobilities in germanium are higher than in silicon. However, the material with the larger energy gap is superior in some applications (see Section 14.9).

The most important elements used for doping silicon and germanium to give extrinsic behaviour are the Group V elements, phosphorus, arsenic and antimony, for n-type and the Group III elements, boron, aluminium, gallium and indium for p-type. As stated in Section 13.6, the room temperature conductivity is a direct function of the amount of doping. The energies of the donor and acceptor states were quoted in Section 13.4 as having specific values in a particular parent material, but there are slight variations for different impurity elements due to differences in their inner electron structures.

Other elements with different numbers of valence electrons can also be used for doping. Thus zinc can accept two and copper can accept three electrons if substituted in the semiconductor material and there will be respectively two and three localized levels for each substituted atom as shown on Figure 14.1.

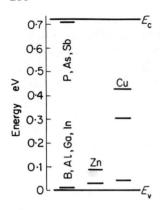

Figure 14.1. Location of some impurity levels in energy gap of germanium. For impurities other than those of Groups III and V there may be more than one level for each atom

Because of the difference in the nuclear charge, these levels are much deeper in the forbidden band than the acceptor levels for Group III elements. The bonding is not entirely covalent, but partly ionic, so that the energies of these levels cannot be calculated simply, but must be found by experiment. The deep levels provide efficient centres for trapping and recombination (see Section 14.3) and so minute traces of these elements as impurities can reduce drastically the minority carrier lifetime.

The second allotropic form of carbon, graphite, is a highly anisotropic material, the conductivity in a direction along the planes of the hexagon structure being a thousand times that perpendicular to the planes. The energy band distribution is such that it is metallic in some directions and semiconducting in others.

The Group VI elements, selenium and tellurium, are semiconductors. Crystalline selenium has an energy gap of 1·8 eV and very low mobilities, $\sim 10^{-4}\,\text{m}^2\,\text{V}^{-1}\,\text{s}^{-1}$. Intrinsic tellurium is an n-type semiconductor with $E_g = 0\cdot33\,\text{eV}$.

14.1.2 Intermetallic compounds

Silicon carbide, which is a compound of silicon and carbon atoms in equal numbers, exists in two polymorphic forms, α and β, which have hexagonal and diamond structures, respectively. Both of these are semiconductors, with energy gaps of 3·1 and 2·2 eV, respectively. Intrinsic conductivity does not begin until about 500 °C. Aluminium and nitrogen are used as doping elements to give p- and n-type behaviour. There is a striking difference in the ionization potentials, these being 0·08 eV for the donor levels and 0·25 eV for the acceptor levels. Hence in the p-type semiconductors, the acceptors are not all ionized at room temperature and there is a significant temperature variation of the properties around this temperature.

Semiconductor Materials and Devices

Compounds made from stoichiometric ratios of the elements which lie to either side of the Group IV semiconductors in the periodic table, e.g. AlP and GaAs, have a diamond structure, with the two kinds of atoms alternating. Thus each Group III atom is surrounded by four Group V atoms and vice versa. They have semiconducting properties and the various combinations provide a wide selection of gap widths and mobilities, some of which are listed in Table 14.1. The main difficulty in preparing these materials is to get the stoichiometric ratio correct to the required accuracy. Some of these have special features, for example indium arsenide (InAs) and indium antimonide (InSb) have very high electron mobilities.

Group II–VI compounds might also be expected to form similar structures with semiconducting properties. Although in compounds such as CdS, the average valency is four, there is a tendency for ionic bonding, the energy gap is greater and the compounds are almost insulators. This tendency is more pronounced as elements are further removed from Group IV in the periodic table. Group I–VII compounds, the alkali halides, where the bonding is almost entirely ionic, have energy gaps in excess of 5 eV.

There are also many ternary compounds, such as $AgSbS_2$ and $CuSbS_2$, which are semiconductors.

14.1.3 Oxides

In simple oxides, such as ZnO and MgO, the bonding is ionic. In the former, the $2p$ levels of the oxygen atoms are filled, forming the valence band, while the empty $4s$ levels of the zinc atoms form the conduction band. The energy gap is 3·3 eV. This compound usually shows n-type semiconductivity attributed to excess zinc atoms, which occupy interstitial positions, acting as donors. Magnesium oxide has a much larger energy gap and is an insulator.

In the oxides of the transition metals, the energy bands which are important for semiconductivity are related to the inner electron levels of the atoms rather than to the outer valence levels. Filled and empty bands from $2p$ anion and $4s$ cation levels are present as in zinc oxide. The $3d$ levels of the cations are only partly filled, so that if the wave functions of $3d$ electrons on adjacent ions overlap sufficiently to form a band, metallic conduction can occur. If no overlap occurs, then as far as $3d$ electrons are concerned, the material is an insulator. In NiO and Fe_2O_3, the wave functions do not overlap, but the forbidden bands are narrow enough for the materials to show semiconducting properties.

14.1.4 Organic materials

Various organic compounds exhibit semiconductor behaviour. While study of these is still in the experimental stage, they show interesting prospects because they may be easier to purify to the required degree and also may be much cheaper than the elemental semiconductor materials.

14.2 Purification of semiconductor materials

The need for high purity in a semiconductor material to give intrinsic behaviour has already been mentioned in Section 14.1. In most semiconductor devices, either p- or n-type material, or both, is required, but otherwise the material must be chemically pure to a high degree. Also the material in the device should not have grain boundaries or other crystallographic defects in the region of any p–n junction. Hence a high degree of crystallographic perfection is also necessary.

The usual method of purification is *zone refining*. A specimen of the material in the form of a bar or wire contained in a suitable trough-shaped crucible, is pulled through a furnace so that a small length of the material is molten at any one time. When liquid and solid are in contact, most impurities will, at equilibrium, have a greater concentration in the liquid than in the solid phase. Hence as the molten zone passes along the specimen, the solidifying material will always be purer than the melt. Each successive pass will increase the degree of purity at the starting end and the impurities will be carried to the finishing end, which can be discarded.*

Zone melting can be used to grow single crystals of any desired crystallographic orientation in the following manner. A *seed* consisting of a single crystal is placed in the desired orientation at one end of the crucible containing the refined material. A molten zone is brought up so that it wets the seed without appreciably melting it and then moved away at a rate slow enough to allow the melt to solidify as a single crystal.

Uniformly doped specimens can be prepared by *zone levelling*. After zone refining, a large concentration of the desired impurity is added to the molten zone at the beginning of a pass. Then, as the doped zone moves along the crystal, the impurity becomes uniformly distributed.

Another method of single-crystal manufacture is the *Czochralski method*. A previously-oriented seed crystal is immersed into the surface of the melt and slowly withdrawn. With suitable control of temperatures and pulling speed, large crystals can be prepared. The composition of the melt can be adjusted to give the desired composition of crystal and redoped at various stages to give layers of different types in the crystal.

Since electrical conductivity is a function of the number of carriers present, conductivity measurements are commonly used as a means of specifying degree of purity or of doping.

---

* If the segregation coefficient is not less than unity, i.e. an impurity does not have a greater concentration in the liquid than in the solid phase, as, for example, boron in silicon, then this method of purification is not applicable.

14.3 Minority carrier lifetime

In a semiconductor, the concentrations of electrons and holes at any temperature reach a state of equilibrium given by equations (13.3) and (13.4). If this equilibrium is temporarily disturbed, for example by electron–hole pair *generation* when light photons (with $hv > E_g$) are absorbed in the material, then when the incident radiation ceases, the excess concentration decays back to the equilibrium value by recombination transitions.

The electron and hole concentrations will change by equal amounts, but in an extrinsic conductor, the majority-carrier concentration is much larger at equilibrium and the change in it will be less marked than that in the smaller minority-carrier concentration. In an *n*-type semiconductor, the rate at which the minority carriers disappear is proportional to the product of the electron and hole concentrations. As the former is almost constant, the rate may be taken as equal to p/τ where τ is a time constant. If electron–hole pairs are being generated in an *n*-type semiconductor, for example at a rate g, then the rate of change of hole density is

$$\frac{dp}{dt} = g - \frac{p}{\tau}$$

When the generating action is removed, this becomes

$$\frac{dp}{dt} = -\frac{p}{\tau}$$

for which

$$p = p_0 \, e^{-t/\tau}$$

as long as p is appreciably large compared with the equilibrium hole concentration. The time constant τ is called the *minority carrier lifetime*.

By illuminating with a light pulse and making subsequent conductivity measurements, the value of τ can be found experimentally. In germanium, for example, τ is found to be less than 10^{-2} s. This value does not agree with the results of wave-mechanical calculations for direct band-to-band recombination which gives τ to be between 10^{-1} and 1 s. It is more likely that the electron returns to the valence band by a succession of transitions as shown in Figure 14.2. A discrete level captures first an electron and then a hole. It has been stated in Section 14.1 that certain impurity elements can give such deep discrete levels which are known as recombination centres. The smaller energy jump at each stage has a much higher probability of occurrence and would reduce the minority-carrier lifetime to the value found experimentally.

In some intermetallic semiconductions, the minority carrier lifetime is much shorter, e.g. $\sim 10^{-9}$ s in gallium arsenide.

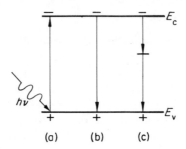

Figure 14.2. Transfer electron across energy gap: (**a**) electron–hole pair generation by a photon, (**b**) direct band-to-band recombination, (**c**) recombination via a discrete level

14.4 Surface states

The periodic potential and the energy bands of the crystal bulk terminate at crystal surfaces. Also any foreign atoms or oxide layer on the surface can give rise to localized quantum states at the surface, called *surface states*, and these influence the properties of the solid in the region near the surface.

Surface states on germanium were postulated by Bardeen to account for various results although the exact character of these states was not specified. In an *n*-type semiconductor, for example, the electrons which occupy these surface states would be taken from the interior and trapped in sufficient number to make the Fermi level in the surface equal to that in the interior. At 0 K, these states would be full up to the Fermi level. If there are higher, unfilled levels, then electrical conductivity along the surface would be possible.

These electrons repel other electrons from the near-surface region and leave positively-charged ionized donors which neutralize the effect of the surface charges. This space charge creates an electric field near the surface and the energy bands will bend upwards, as shown in Figure 14.3, because the potential energy of an electron would be higher in this region.

If the bending causes the energy levels to shift relative to the Fermi level until the level is below the middle of the band, then the near-surface region will be *p*-type and holes in the valence band will become the majority carriers and assist the ionized donors in neutralizing the surface charges. The material is then said to have an *inversion* layer at the surface as is shown in Figure 14.4. If the inversion layer forms, then a very large number of valence states are available so that no further bending is necessary even for very high surface charge densities. The maximum height of the band above the Fermi level is thus equal to

$$E_g - (E_c - E_0)$$

Semiconductor Materials and Devices

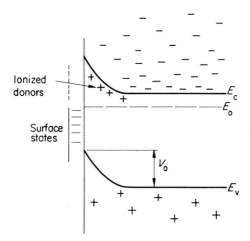

Figure 14.3. Charged surface states which induce a surface potential barrier

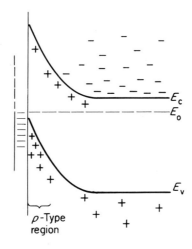

Figure 14.4. Strongly charged surface states produce a p-type inversion layer on an n-type crystal

The rise in the band at the surface provides a potential barrier to the free movement of electrons between the surface and the interior. Because it is asymmetrical, it has a rectifying action as is explained in Section 14.6 for the case of a metal–semiconductor rectifying contact.

14.5 Contacts

When two otherwise insulated pieces of conducting or semiconducting materials are placed in contact, there will be a transfer of electrons from one to the other unless the Fermi energies happen to be identical (Figure 14.5). As

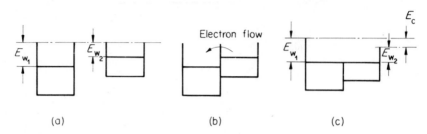

Figure 14.5. Two metals (**a**) before contact, (**b**) immediately after contact, (**c**) after attainment of equilibrium

this transfer proceeds, the potentials of the materials change. The one to which the electrons flow becomes negatively charged and its energy levels rise, while the other becomes positively charged and its energy levels fall. Equilibrium is attained when the Fermi levels are equal, the net electron flow then being zero.

In the case of two metals, the contact potential difference E_c between them is the difference in the work functions

$$E_c = E_{w_1} - E_{w_2}$$

Because the work functions for most metals lie in the range 1–10 V, E_c is usually of the order of 0·1–1 V.

The Fermi level of a metal falls with increasing temperature as described in Section 10.6. If the rate of fall differs in the two metals, then the potential difference E_c varies with temperature. This is the source of the Seebeck potential described in Section 11.8.

In a semiconductor–metal contact, two possibilities occur depending upon which has the higher Fermi level. These two cases are considered in the subsequent sections.

14.6 Rectifying contacts

If the Fermi level of the metal is below that of an *n*-type semiconductor, then upon contact, electrons will flow from the semiconductor to the metal until the Fermi levels are equal. The extra negative charges then in the metal repel free electrons from the surface of the semiconductor, leaving positively-charged ionized donors to neutralize the effect of the negative charges. This causes an upward bending of the energy bands near the surface (Figure 14.6) similar to

Semiconductor Materials and Devices

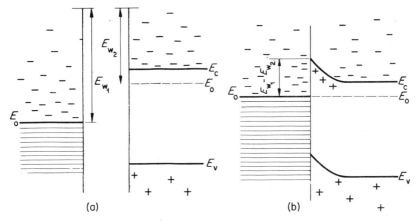

Figure 14.6. Energy level diagram of junction between a metal and an *n*-type semiconductor when the work function of the metal is larger than that of the semiconductor: **(a)** before contact showing thermally-excited electrons in metal, **(b)** after contact showing space charge layer

that described in Section 14.4. In the bulk of the semiconductor, electron–hole pairs will form to keep the electron–hole product, *np*, constant. Hence the position of the Fermi level relative to E_c and E_v will not change.

An analogous case occurs when the semiconductor is *p*-type and when the work function of the metal is smaller (i.e. its Fermi level is higher). Then electrons flow from the metal to the semiconductor giving the situation shown in Figure 14.7.

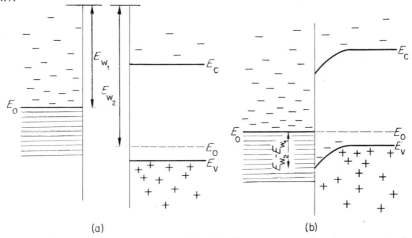

Figure 14.7. Energy level diagram of junction between a metal and a *p*-type semiconductor when the work function of the metal is less than that of the semiconductor: **(a)** before contact, **(b)** after contact

In both cases a potential barrier forms at the junction, the height of the barrier equalling the difference of the work functions. In the former case, the barrier impedes the flow of electrons and in the latter the flow of holes.

When a potential difference V is applied across the junction, the metal, being a good conductor, remains at constant potential and the gradient is entirely in the semiconductor (Figure 14.8). The current flow from the metal is determined

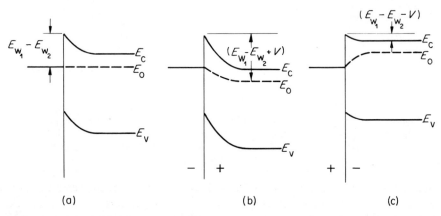

Figure 14.8. Effect of applied potential difference V at metal–n-type semiconductor junction: (a) zero bias, (b) reverse bias, (c) forward bias

by the thermionic emission of electrons (see Section 11.6)

$$I_1 = BA_0 T^2 \, e^{-(E_{w_1} - E_{w_2})/kT}$$

where B is the area of contact and A_0 is a constant.

The current from the semiconductor is

$$I_2 = BA_0 T^2 \, e^{-(E_{w_1} - E_{w_2} - eV)/kT}$$

where the sign of V is taken to be the polarity of the metal. The net current is

$$I = I_2 - I_1$$
$$= I_s(e^{eV/kT} - 1)$$

where $I_s = I_1$ is the saturation current.

When V is negative and increases, the reverse current rises to the constant value I_s and when V is positive and increases, the forward current increases exponentially, the complete I–V relation being shown in Figure 14.9.

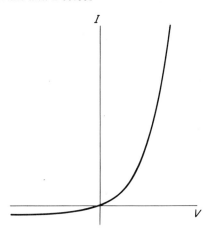

Figure 14.9. Current–voltage relationship for rectifying junction

Hence, while rectification is not complete, the large forward-to-reverse current ratio of rectifying contacts can be utilized for rectification purposes. In such devices, the rectifier contact is usually in the form of a metal point resting on the surface of the semiconductor while the other electrode is a soldered ohmic contact (see next section).

In the point-contact germanium rectifier, a whisker of a metal such as platinum, gold or tungsten is held with a pointed end in firm contact with the surface of a germanium crystal—usually n-type. These are of small size and are used for low-current rectifying purposes in electronic circuits. With low resistivity material, the rectifier is of the metal–semiconductor type and the rectification depends upon the whisker material. With high resistivity material it is more likely to behave as a surface state-type rectifier and the characteristics are independent of the whisker material.

14.7 Ohmic contacts

If the Fermi level of an n-type semiconductor is below that of the metal, electron flow makes the surface of the semiconductor more n-type and the energy bands are bent downwards as in Figure 14.10. There is no potential barrier to electron flow between the solids. The p-type junction will behave similarly with the energy bands bending upwards if the Fermi level is above that of the metal.

Hence the presence or otherwise of a potential barrier depends upon the relative values of the work functions. If, however, there are surface states on the semiconductor (as described in Section 14.4), these may be more important in

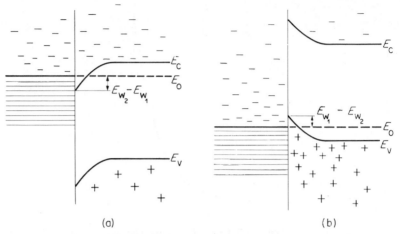

Figure 14.10. Energy level diagram of metal–semiconductor junctions for ohmic contacts: (**a**) *n*-type, (**b**) *p*-type

forming and determining the height of the barrier. To get ohmic contacts, which may be necessary when any rectifying action must be avoided, it is necessary to remove as many of the surface states as possible before applying the metal contact.

14.8 The *p–n* junction

A *p–n* junction consists of a single crystal of semiconductor in which there is an excess of donors in one part and an excess of acceptors in another. If the composition between the two regions changes gradually, it is a *graded* junction, whereas if it is abrupt it is a *step* junction.

When the two regions are initially in contact and isolated from any circuit, the concentration of electrons at higher levels which is much larger in the *n*-type region, will cause a net electron current from the *n*- to the *p*-type region and a hole current in the reverse direction. These change the relative potentials of the regions so that the energy levels shift until the Fermi levels coincide. In an isolated *p–n* junction, electrons and holes can flow in either direction due to thermal agitation.

In Figure 14.11, I_1 is the electron current from the *n* to the *p* region and is due to the electrons in the *n* region which have energies above the E_c value of the *p* region. The current I_2 in the reverse direction is due to electrons in the *p* region near the junction and at the bottom of the conduction band 'falling over' the potential 'hill' into lower vacant energy levels in the *n* region.

In an isolated junction, these two currents must eventually be equal, but this equilibrium condition is upset if electrodes are connected to the two sides of the junction and maintained at different potentials as in Figure 14.12. The external

Semiconductor Materials and Devices

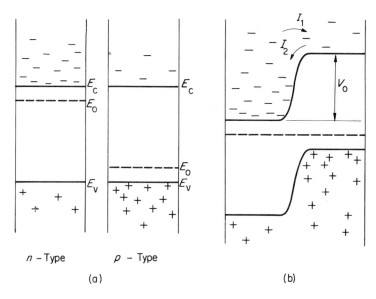

Figure 14.11. *n*- and *p*-type semiconductors: (**a**) before contact, (**b**) after contact and equilibrium is attained

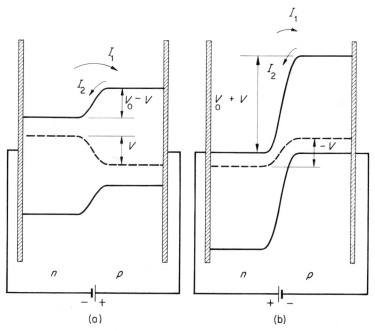

Figure 14.12. Effect of biasing a *p–n* junction: (**a**) forward bias increases I_1, (**b**) reverse bias decreases I_1; I_2 is constant

connections should be made through ohmic contacts to avoid any effect due to potential barriers at the semiconductor–electrode interface.

When the n-type is made negative and the p-type positive, the bias increases the height of the n region energy bands relative to those in the p region so that current I_1 increases. The reverse current I_2 is unchanged, giving a net current in the forward direction.

The application of a bias voltage V in the reverse direction reduces I_1 while I_2 is again unchanged.

I_2 is proportional to the equilibrium electron concentration n_p in the p region

$$I_2 = C_1 n_p$$

C_1 is a constant that involves the area of the junction and parameters of the semiconductor.

I_1 is proportional to the number of electrons in the n region with sufficient energy to surmount the barrier

$$I_1 = C_1 n_n \, e^{-(V_0 - V)/kT}$$

where n_n is the electron concentration in the n region.

When $V = 0$, $I_1 = I_2$, so that

$$n_p = n_n \, e^{-V_0/kT}$$

When a bias voltage is applied, the net electron current is

$$I_n = I_1 - I_2$$
$$= C_1 n_n (e^{V/kT} - 1) e^{-V_0/kT}$$
$$= C_1 n_p (e^{V/kT} - 1) \tag{14.1}$$

The hole current is given by a similar expression

$$I_p = C_2 p_n (e^{V/kT} - 1) \tag{14.2}$$

where p_n is the equilibrium concentration of holes in the n region. The total current is the sum of equations (14.1) and (14.2), i.e.

$$I = I_n + I_p$$
$$= I_0 (e^{V/kT} - 1)$$

where

$$I_0 = C_1 p_n + C_2 n_p$$

is the total current flowing in either direction when $V = 0$ and is called the *saturation current*. The current I varies with V in precisely the same manner as for metal–semiconductor contacts.

Semiconductor Materials and Devices

14.9 Applications of p–n junctions

14.9.1 *Junction rectifier*

The action described in the preceding section can be utilized for rectifying alternating currents. Complete rectification is not achieved, because of the saturation current that flows when reverse bias is applied. This can be kept small by reducing n_p and p_n, i.e. the number of minority carriers in each part of the junction, or by using a material with a wider forbidden energy band. The saturation current can then be of negligible importance as may be seen from a typical device characteristic shown in Figure 14.13.

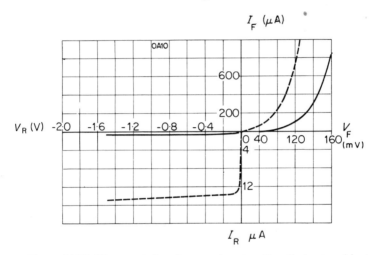

Figure 14.13. Characteristics of germanium junction diode at ambient temperatures of 25 °C (full line) and 60 °C (broken line). Note the change at the origin in the scales of both axes. (Acknowledgements to Mullard Ltd.)

It will be seen that the rectification efficiency decreases with increasing temperature. Because the saturation current is increased by increasing the temperature, a material with a wider energy gap is more essential if the heating due to electrical losses or any other effect causes a temperature rise. Hence silicon and silicon carbide are in many cases superior to germanium as rectifier materials.

When the junction is forward biased, holes travel from the *p*- to the *n*-type region and electrons from the *n*- to the *p*-type region. This increases the minority carrier population in each region, so that recombination occurs. The energy so released is evolved mostly as heat in silicon and germanium, but in gallium arsenide and some other semiconductors it appears as light, the photons emitted having approximately the energy of the gap.

14.9.2 Zener diode

When a high reverse bias is applied to a *p–n* junction, the large electrical field across the junction accelerates the carriers to such high velocities that they cause ionizations by collisions. The electrons so produced may also be accelerated sufficiently to cause further ionization, so giving an avalanche breakdown. The typical forms of the forward and reverse current characteristics are shown in Figure 14.14.

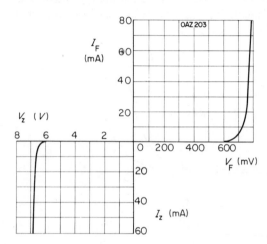

Figure 14.14. Forward and reverse characteristics of zener diode. Note the change in the scale of the voltage axis. (Acknowledgements to Mullard Ltd.)

Unless the power dissipated is so large as to cause local melting, this avalanche breakdown is not destructive and the properties of the junction are not permanently affected.

During the breakdown, the voltage across the junction remains fairly constant over a large range of values of current—so that this effect can be used for a constant voltage device. Junctions designed specially for use in this manner are known as *Zener diodes*.

14.9.3 Photocells

Photons falling on a semiconductor will generate electron–hole pairs if $h\nu$ is greater than E_g. Any carriers generated at the junction and any generated nearby which diffuse to the junction will be moved by the electric field of the junction, holes to the *p* side and electrons to the *n* side because vacant levels for them exist (see Figure 14.15). If the device is isolated, a change in the potential will build up until there is an equal reverse current to preserve equilibrium.

If the device is connected to an external circuit, then a current will flow, the magnitude of which is proportional to the intensity of the incident illumination.

This is the principle of the photoelectric cell used both for the detection of light and for the measurement of the intensity of illumination.

Semiconductor Materials and Devices

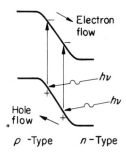

Figure 14.15. Creation of electron–hole pairs at p–n junction by action of photons

The energy gap in silicon corresponds to a wavelength of 10 780 Å and so a silicon junction will be sensitive to all of the visible spectrum. As may be seen from Table 14.1, the energy gaps of the Group VI compounds of cadmium

Table 14.1 Properties of some semiconductors

Material	Energy gap eV	M.P. °C	ε_r	Mobilities $m^2 V^{-1} s^{-1}$	
				μ_n	μ_p
Si	1·1	1417	11·7	0·13	0·05
Ge	0·67	937	16·3	0·39	0·19
β Se	1·74	220	~ 8·5		0·0006
Te	0·33	450	2·2–5·0^a	0·11	0·50
SiC	2·8	2830	10·0	0·04	0·005
AlP	3·0	>1500	11·6	0·008	
AlAs	2·16	>1600	8·5	0·018	
AlSb	1·6	1054	10·1	0·002	0·0004
GaP	2·24	1350	8·5	0·01	0·0075
GaAs	1·35	1237	10·4	0·85	0·04
GaSb	0·7	712	14·0	0·4	0·14
InP	1·27	1054	12·1	0·4	0·065
InAs	0·36	942	11·7	3·3	0·046
InSb	0·18	525	15·6	8·0	0·075
Cd0	2·5	>1420		0·012	
CdS	2·6	1750	5·4	0·034	0·0018
CdSe	1·74	1350	10·0	0·06	
CdTe	1·44	1098	11·0	0·03	0·0065
ZnO	3·2	1975	8.14	0·018	
PbS	0·38	1077	17·0	0·06	0·02
PbSe	0·26	1062	23·6	0·14	0·14
PbTe	0·28	904	30·0	0·60	0·40

^{a} Anisotropic.

Extracted from '*Semiconductors*' by H. F. Wolf, Wiley-Interscience, 1971 and other sources.

The values quoted may vary with temperature and in the case of ε_r with frequency.

correspond to the energies of photons in the visible or near visible region and can be used to detect at least some of the visible region. The spectral response characteristic of cadmium sulphide is shown in Figure 14.16.

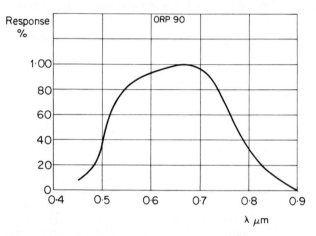

Figure 14.16. Spectral response curve of cadmium sulphide photocell. (Acknowledgements to Mullard Ltd.)

To detect infrared radiation, semiconductors with smaller energy gaps are needed. Germanium with an energy gap corresponding to 17 200 Å can be used for the near infrared region and indium sulphide with E_g equivalent to 77 500 Å will detect radiation in the far infrared region.

14.9.4 *Solar batteries*

The *photovoltaic effect* described in the previous subsection can be utilized to generate electricity from solar energy on a useful scale. Silicon junctions are mostly used for this purpose because the energy gap is appropriate.

The carriers must be produced at or near the junction and so this must be near the surface or the incident radiation would be absorbed before reaching it. The usual form of construction is an *n*-type crystal with a *p*-type layer of about 10^{-6} m thickness at the surface. A single cell can give an output of about 0·5 V and the best overall efficiencies that can be achieved for energy conversion is about 15 per cent.

14.9.5 *Tunnel diodes*

At heavy doping levels of 10^{25} m^{-3} or more, the electron or hole concentration is degenerate and the Fermi level lies in the conduction or valence band (Figure 14.17). With very narrow junctions (~ 100 Å) the wave function of an electron in a state of the conduction band of the *n*-type region extends into

Semiconductor Materials and Devices 233

states in the valence band of the *p*-type region and vice versa. Therefore there is a finite probability that an electron can cross the junction through the potential barrier by tunnelling in the manner described in Section 3.1.2. The tunnelling current each way depends upon the junction width, the number of electrons which are capable of tunnelling and the number of empty states into which they can move. At equilibrium the two currents, one each way, must be equal.

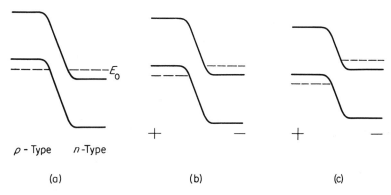

Figure 14.17. Relative movement of energy bands on biasing tunnel diode: (a) unbiased, (b) and (c) increasing forward bias

When a small forward bias is applied (Figure 14.17(b)) the occupied states of the *n*-type conduction band become level with the empty states of the *p*-type valence band so that a large number of electrons tunnel from *n* to *p* giving a large current in the reverse direction.

With still further increase in bias, so that the conduction band of the *n*-type does not overlap the valence band of the *p*-type region, there can be no tunnelling and the only current will be that normal for a forward biased rectifying *p–n* junction. Hence the effect of increasing bias is firstly an increase of current from zero and then between the biases appropriate to Figures 14.17(b) and (c), a decrease of current. In this latter range, the junction appears to have a negative resistance.

Devices operating on this principle are known as tunnel diodes and have current–voltage characteristics of which that shown in Figure 14.18 is typical.

Because the tunnelling effect is not limited by carrier diffusion, tunnel diodes can operate as much as a hundred times faster than transistors. For this reason tunnel diodes can be used to advantage in certain applications such as computer switching circuits.

Silicon and germanium are commonly used for these devices. Gallium arsenide has also been used and has the advantage that it can operate at temperatures approaching 350 °C.

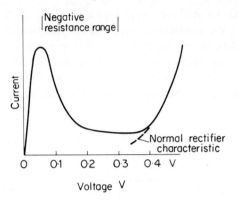

Figure 14.18. Typical characteristic of a tunnel diode

14.9.6 *Capacitance*

The *p–n* junction is a double layer of opposite charges separated by a small distance and so has the properties of an electrical capacitance.

As reverse bias is applied, majority carriers are attracted away from the junction and the depletion zone is widened so that the capacitance decreases. Little direct current flows when the junction is biased in this direction, but at high frequencies, the capacitance can pass appreciable alternating currents without rectifying action. By varying the biasing d.c. voltage, the capacitance can be changed. Devices designed specifically for this application are known as *voltage-variable capacitance diodes*.

14.10 The junction transistor

The junction transistor is a single crystal of semiconducting material containing two *p–n* junctions which are separated by a distance less than the minority carrier diffusion length. The transistor may be either a *n–p–n* or a *p–n–p* type and the two types are identical in their mode of action except for the interchange of the two sorts of carrier and the change of polarity of the bias voltages. The *n–p–n* type is described here. The two *n*-type regions are known as the *emitter* and *collector* and the *p*-type region as the *base*. A conducting lead is attached to each. Under equilibrium conditions, the Fermi level will be constant throughout so that the energy bands are as shown in Figure 14.19(a). To operate as an amplifier, the emitter is given forward bias and the collector reverse bias as shown in Figure 14.19(b).

Of the electrons in the emitter with sufficient energy and travelling in the correct direction (to the right in Figure 14.19), the majority will diffuse through the base region to the collector where there are lower energy levels which they

Semiconductor Materials and Devices

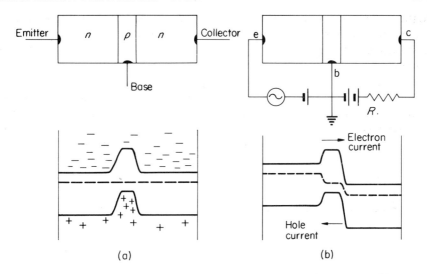

Figure 14.19. *n–p–n* junction transistor: **(a)** unbiased, **(b)** biased as amplifier

can occupy. Flow from the collector in the reverse direction will be negligible. If the value of the emitter–base voltage is changed, then the number of electrons that have sufficient energy to overcome the potential barrier of the base will also change and the collector current varies. If an alternating signal is applied to the base, then there will be an alternating voltage observed across the load resistance R. Since there is very little current and hence little power input to the base, the transistor operates as a power amplifier.

The requirements for maximum amplification are:

a. The electron current from the emitter should be large, so therefore the emitter region should be heavily doped.
b. Most of the emitter current electrons should diffuse across the base region, so that the minority carrier lifetime and mobility in this region should be high. Hence it should be lightly doped.
c. For the same reason, the base region should be thin.
d. The carrier density in the collector region is not so important, but by making this part lightly doped, the capacitance of the collector junction is reduced. Also a higher resistivity of this region gives better amplification.

14.11 The field-effect transistor (FET)

This device consists of a *channel* of semiconductor material with two *p–n* junctions made on opposite sides of the channel. The channel may be of either *n*- or *p*-type material. Conducting leads are connected through ohmic contacts to the two ends of the channel. In operation, an external voltage is applied across these contacts so that majority carriers are injected at the *source* and collected at the *drain*. The magnitude of the current due to these majority carriers depends upon the conductivity of the channel and the source–drain voltage.

The portions of the material on the other sides of the *p–n* junctions, i.e. the *p* regions in an *n*-channel device and vice versa, are connected electrically and are known as the *gate*. When the *p–n* junctions are reverse biased, majority carriers are removed from the immediately adjacent regions of the channel. These regions are known as *depletion* layers. Because there are few free carriers in a depletion region, the resistivity is high. The thickness of the depletion layers and hence the effective conducting area of the channel vary according to the magnitude of the reverse voltage.

A field-effect transistor connected to voltage sources is shown in Figure 14.20.

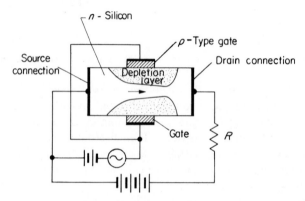

Figure 14.20. Schematic diagram of field-effect transistor in use as an amplifier

Because there is a voltage drop along the length of the channel, the reverse bias across the *p–n* junction will also vary along the length, being greatest at the drain end. The depletion regions will therefore be wedge-shaped as shown.

The special features of the field-effect transistor compared with the junction transistor are: a very high input resistance, conduction by majority carriers, low noise, low drift and good frequency response. These features make it possible by replacing junction transistors with field-effect transistors to improve circuit performance and indeed design other circuits not possible with junction transistors.

14.12 The metal-oxide semiconductor transistor (MOST)

This is another type of field-effect transistor in which a conducting channel is induced between two closely-spaced electrode regions in a semiconductor by increasing the electric field at the surface of the semiconductor between the two regions.

In a p-channel MOST, the source and drain are two high-concentration p-type regions diffused into an n-type silicon wafer and separated by a very small distance (Figure 14.21). A thin layer of very pure silicon dioxide, about 1500 Å

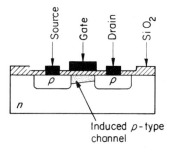

Figure 14.21. Basic structure of a metal-oxide semiconductor transistor

thick, is formed on the silicon surface between the source and drain and a metal electrode, the gate, deposited on the top of this layer. When the gate is made sufficiently negative with respect to the source, holes will be attracted to the surface of the n-type material under the electrode and so form a p-type channel. The gate voltage at which conduction starts is called the *threshold voltage*. Because the gate electrode is separated from the rest by an insulating layer of silicon dioxide, the input impedance is extremely high.

The method of manufacture of these devices follows the steps described generally in Section 14.14.

14.13 Other semiconductor applications

14.13.1 *Hall effect devices*

Semiconductor specimens can be used to measure the strength of magnetic fields. A known current I_c is passed in one direction through a crystal placed in the field and the Hall voltage is measured (Figure 14.22).

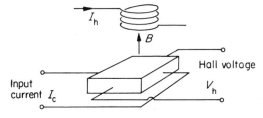

Figure 14.22. Use of Hall effect for measurement of magnetic field strength

If the magnetic field is set up by another current I_h flowing in a solenoid, then the flux density at the crystal is proportional to I_h and the Hall voltage is proportional to the product of the currents

$$V_h = CI_c I_h$$

This principle can be used as a multiplying device in an analogue computer and also as a wattmeter, in which case the line voltage would be applied to the field coils and the line current passed through the crystal. The output would then be read on a d.c. meter.

To get accurate results in such applications, it is essential that the temperature of the crystal be controlled within narrow limits. To get large outputs, materials with high carrier mobilities such as indium arsenide and indium antimonide are needed.

14.13.2 Thermistors

The temperature dependence of the conductivity of semiconductors can be used in applications for temperature measurement. Thermally sensitive resistors made for this purpose are known as *thermistors*. The conductivity variation with temperature for some materials is so great that it is possible to record changes of temperature as small as 10^{-6} K.

Also because the conductivity increases with temperature, whereas that of metals decreases, thermistors can be used in circuits to compensate for conductivity changes and so achieve systems with zero temperature coefficients.

These devices are usually made from sintered oxides, a mixture of several oxides being used and the proportions varied to give different characteristics. A typical temperature characteristic is shown in Figure 14.23 from which the large range is readily apparent.

Because conductivity should vary as $e^{E_g/2kT}$, then the resistance of a thermistor with an energy gap of 0.3 eV would vary by about 4 per cent. for a temperature change of one degree at room temperature.

When a current is passing through a thermistor, heating takes place thereby altering the temperature and decreasing the resistance. When sufficient time is allowed for an equilibrium state at any current to be attained, the current–voltage relationship is of the type shown in Figure 14.24. The curve is ohmic for small currents, passes through a maximum and then has a negative resistance characteristic. As such, it finds applications in many circuits, but it is limited to low frequency variations because of the time required to attain thermal equilibrium.

The exact *V–I* characteristic depends upon the rate at which the thermistor can lose heat and this is a function of the heat-transfer properties of the surroundings. In a gas, the heat-loss rate depends upon the pressure and velocity

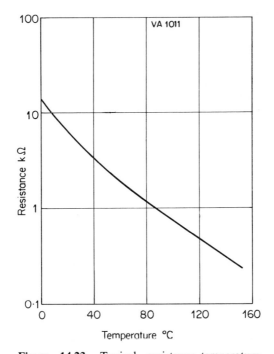

Figure 14.23. Typical resistance–temperature characteristic of a thermistor. (Acknowledgements to Mullard Ltd.)

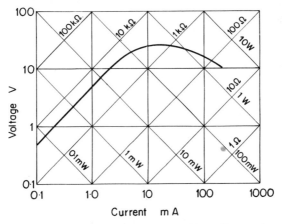

Figure 14.24. Steady-state current–voltage characteristic of a thermistor for ambient temperature of 25 °C. (Acknowledgements to Mullard Ltd.)

and hence thermistors can be used as the sensing elements in pressure gauges, flow meters, etc.

Another application is for time-delay switching. For a thermistor with a characteristic like Figure 14.24, there will be a delay between switching on the current and it building up to a steady value. Delays from milliseconds to minutes can be obtained by variations in design.

14.13.3 *Varistors*

Devices with non-linear voltage–current relationships, which are almost independent of the direction of current flow, are known as *varistors*.

Silicon carbide is one material with such a property and is used for applications like lightning arrestors. The characteristic is of the form

$$I = kV^\alpha$$

where α usually lies between four and six, the actual value depending upon the exact composition and treatment of the material.

Such devices are shunted across the equipment which is to be protected. They pass very little current at normal operating voltages, but pass large currents when subjected to voltage surges, such as those due to lightning strikes.

14.14 Microminiature solid state circuits

The diodes and transistors described in Sections 14.9 to 14.12 are usually manufactured from a starting point of either *p*- or *n*-type single crystal material. By placing the appropriate impurity on to the surface and maintaining a high temperature for sufficient time for solid-state diffusion to occur, regions of the other type are grown.

The technique has been further extended to the stage where several devices with the necessary interconnections can be fabricated on a single crystal to give a complete circuit. These circuits occupy much less space than would be the case if separate devices wired together were used and it is possible to get packings as high as 180 components per cm^3.

The principle can be illustrated by first describing the planar technique of manufacturing silicon transistors. A typical starting point might be a slice of *n*-type silicon about 0·04 m diameter and 0·1–0·2 mm thick on which perhaps 1000 transistors can be formed.

An oxide layer of the order of 1 μm thick is grown on to the surface of the slice under scrupulously clean conditions (Figure 14.25). By a photolitho process, the areas which will become the base regions of the transistors are defined and the oxide film in these areas is etched to expose the unoxidized silicon. The slice is placed in a boron-rich atmosphere and boron diffuses into the silicon in the exposed regions to form *p*-type layers, the silicon dioxide masking the silicon

Semiconductor Materials and Devices

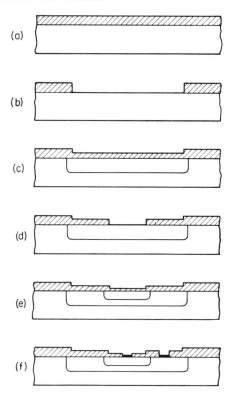

Figure 14.25. Stages in manufacture of planar transistor. (**a**) SiO$_2$ layer grown on n-type silicon slice, (**b**) window etched in oxide layer, (**c**) base region and further oxide layer grown, (**d**) window etched, (**e**) emitter region and oxide layer grown, (**f**) metallic contacts added

elsewhere. The boron diffuses laterally as well as into the depth so that the p–n junction at the silicon surface is under the oxide layer. This junction, if exposed, would have an electric field across it and so attract any ionic material. Because it is covered, contaminants cannot reach it and this prevents unwanted surface states. During the boron diffusion, a further oxide film grows which is also treated to expose smaller windows that will define the emitter regions. Exposure to a phosphorus-rich atmosphere gives the necessary diffusion to form n-type regions and finally a complete silicon dioxide covering. Windows are etched in the appropriate places in the oxide film to enable metallic contacts to be made to the appropriate areas of the transistor.

Such a slice can then be cut into pieces, each of which contains one transistor, which is then suitably encapsulated for use.

If a group of transistors were kept intact, they could be used in a circuit in which their collectors were required to be in electrical connection. By depositing metallic leads over the insulating oxide layer between the external contacts, other circuit connections can be made as illustrated in Figure 14.26.

Figure 14.26. Two transistors on a single slice with emitter of one connected to base of other

Components can be isolated from each other if they are grown entirely as diffused layers and the first layer is given a potential such that a reverse biased p–n junction is formed between it and the substrate as shown in Figure 14.27.

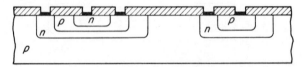

Figure 14.27. Transistor and diode grown in p-type substrate which can be isolated by applying a reverse bias to the p–n junctions

Resistors are made by having direct paths of semiconductor material between components or by diffusing heavily doped channels in the substrate. With suitable polarity to give reversed bias junctions, these can be insulated from the substrate (Figure 14.28(a)).

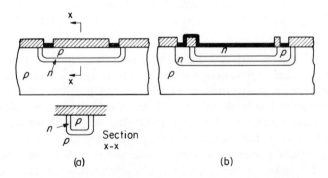

Figure 14.28. Isolated (**a**) resistor, and (**b**) junction capacitor

Semiconductor Materials and Devices

Capacitors can be obtained by using the capacitance of a reversed bias junction as in Figure 14.28(b), the capacitance being proportional to the area involved.

An important and laborious stage in the manufacture of these solid-state circuits is the preparation of the masks for defining the areas to be etched in the diffusion layers for each of the diffusion stages and for defining the paths of the metallic leads. Hence the method is only suitable where large numbers of any one circuit are required.

Questions for Chapter 14

1. Discuss the difference between n-type and p-type germanium and describe briefly the operation of the junction transistor.

    The electron–hole pair density in a sample of germanium is 10^{22} m^{-3} and the mobilities are $\mu_n = 0.39$ and $\mu_p = 0.19$ m^2 V^{-1} s^{-1}. Calculate the conductivity. [MST]

2. A semiconductor is said to have a density of states N_c m^{-3} for its conduction band and similarly a density of states N_v m^{-3} for its valence band. Explain the concept of density of states for an energy band. Taking the zero energy reference level as the top of the valence band and denoting the Fermi level by E_f and the bottom of the conduction band by E_c, write down an expression for the number of electrons and holes in the conduction band and valence band, respectively, in terms of N_c, N_v, E_f and E_c. Explain any approximations that have been made.

    In addition there are N_d donors m^{-3} at an energy level δE below the conduction band. With the same approximations as made previously write down the numbers of electrons in these donor levels and explain where the remainder have gone. Hence find an expression for the Fermi level, E_f.

    How does the Fermi level vary with temperature? What effect does this variation have on the conductivity? Why does the manner of this variation suggest that silicon may be preferred to germanium for many applications? [E]

3. (a) At room temperature the intrinsic density of mobile carriers in pure germanium is n_i per unit volume both for electrons and for holes, and the Fermi level lies near the centre of the forbidden gap.

    If N_D donors per unit volume are added to the material to make it n-type, the Fermi level will rise by energy ΔE. Derive an expression for ΔE in terms of N_D, n_i, the absolute temperature T and Boltzmann's constant k. It may be assumed that all donors are ionized, that n_i is negligible in comparison with N_D and that the classical approximation to the Fermi function is justified.

(b) In a germanium p–n junction the p-type material has resistivity 10^{-2} Ωm, while the resistivity of the n-type is 10^{-3} Ωm. The mobilities of holes and electrons are respectively

$$\mu_p = 0.19 \text{ m}^2 \text{ V}^{-1} \text{ s}^{-1} \qquad \mu_n = 0.39 \text{ m}^2 \text{ V}^{-1} \text{ s}^{-1}$$

and the resistivity of intrinsic material is 0.47 Ωm. Assuming all donors and acceptors to be ionized, use the expression derived in the previous section to calculate the built-in voltage step in the junction taking $T = 300$ K. [EST]

4. For a particular intrinsic semiconductor the equilibrium value of the n–p product is 10^{34} m^{-6} and the excess carrier lifetime is 500 μs. Under the action of a light source, electron–hole pairs are generated at a rate of 10^{21} m^{-3} s^{-1}. Calculate the values of n and p and also derive an expression for the change of these values with time after the light is turned off.

5. A gallium arsenide semiconductor has the same structure as diamond with the two kinds of atoms alternating. The lattice constant is 5.43 Å.
Calculate the density of gallium arsenide.
If the stoichiometric ratio is unbalanced to the extent of five arsenic atoms in every 10^6 being replaced by gallium atoms, what will be the number of carriers per m^3? By what factor will the conductivity be increased above that of material with the stoichiometric ratio of atoms? Assume $N_c = N_v = 2.5 \times 10^{25}$ m^{-3} and use values of E_g, μ_n and μ_p from Table 14.1.

6. Explain the action of a p–n–p transistor showing, by means of energy-level diagrams, the relative positions of the Fermi-level, the valence band and the conduction band (a) in the unbiased transistor, and (b) when the usual direct bias voltages have been applied. [EST]

7. In a p–n–p transistor the boundaries of the three regions are parallel planes and it may be assumed that the flow of current is wholly in a direction perpendicular to these planes. The width of the base is d, the diffusion coefficient for holes is D and the diffusion length $L = \sqrt{D\tau}$, where τ is the average lifetime of holes in the base region.
The emitter/base junction is biased in a forward direction and a current I_1 of holes is injected into the base, where the electric field is negligibly small. As a result of recombination, the hole current I_2 which diffuses to the base/collector junction, and is there removed, is somewhat smaller than I_1. The equilibrium concentration of holes in the base region when $I_1 = 0$ is negligibly small compared with the concentration when I_1 is flowing and d is much smaller than L.

Semiconductor Materials and Devices

Show that, under these conditions,

$$\frac{I_2}{I_1} = 1 - \frac{d^2}{2L^2} \qquad \text{[EST]}$$

8. Describe the metal-oxide silicon transistor (MOST) and its mode of operation.

List the steps required in making integrated microcircuits by the planar process. [E]

CHAPTER FIFTEEN

Magnetic Materials

15.1 Basic concepts

In vacuo, a magnetizing force H produces a flux density B, the two being related by μ_0, the magnetic permeability of free space

$$B = \mu_0 H$$

With H measured in A m^{-1} and B in T, μ_0 has the value $4\pi \times 10^{-7}\ \text{H m}^{-1}$.

When a material is placed in a magnetic field of magnetizing force H, it becomes magnetized, the intensity of induced magnetism being M. The flux density external to the medium is the sum of that due to the original magnetizing force and that due to the induced magnetization, these quantities being related by the equation

$$B = \mu_0(H + M)$$

Also

$$B = \mu H$$

where μ is the absolute permeability of the material. The ratio of μ to μ_0 is known as the *relative permeability*

$$\mu_r = \mu/\mu_0$$

Also the magnetic *susceptibility* is defined as

$$\chi_m = \frac{M}{H}$$
$$= \frac{B}{\mu_0 H} - 1$$

so that

$$\chi_m = \mu_r - 1$$

Magnetic Materials

The B–H relationship of a material is not necessarily linear, so that susceptibility may not be constant. The sign and magnitude of χ_m and its manner of variation with H and temperature define the magnetic behaviour of a material and form a basis of classification.

At one time, the classes were recognized as diamagnetic, paramagnetic and ferromagnetic. However, the first two can be subdivided depending upon the temperature variation of their magnetic properties while antiferromagnetic and ferrimagnetic materials are recognized as subclasses of ferromagnetic materials. The various classes are outlined in Table 15.1 and considered in more detail later in the chapter.

Table 15.1 Magnetic susceptibility

Name	Sign	Magnitude	Temperature dependence	Characteristic materials
Diamagnetic	−	Small	Independent	Inert gases, alkali halides, organic compounds
	−	Large $\chi_m = 1$	Independent only below T_c, which is always < 20 K	Superconducting metals
Paramagnetic	+	Small to moderate	Independent	Alkali and transition metals
	+	Moderate to large	$\chi_m = \dfrac{C}{T - \theta}$	Rare earths, gases
Antiferromagnetic	+	Small	$T > T_c, \chi_m = \dfrac{C}{T - \theta}$	Salts of transition metals
			$T < T_c, \chi_m \propto T$	
Ferromagnetic	+	Very large	$T > T_c, \chi_m = \dfrac{C}{T - \theta}$	Some transition and rare earth metals
			$T < T_c$, unique variation	
Ferrimagnetic	+	Very large	$T > T_c, \chi_m = \dfrac{C}{T \pm \theta}$	Ferrites and garnets
			$T < T_c$, qualitatively similar to ferromagnetic but with quantitative differences	

Adapted from Table 12.1 of *The Science of Engineering Materials*, Ed. J. E. Goldman, John Wiley, New York, 1957.

15.2 Atomic magnets

The magnetic properties of materials are associated with the magnetic moments of the individual atoms, which are due to the motion of the electrons in their orbits and to electron spin.

Charges of magnitude q passing a point one at a time at constant time intervals t are equivalent to an average current q/t. Hence an electron describing an orbit about the nucleus is equivalent to a current given by

$$I = -e/\tau$$

where τ is the periodic time.

The angular momentum of the electron remains constant and can be represented by a vector **G** which is perpendicular to the plane of the orbit and is of magnitude $mr^2\dot\theta$, where r and θ are the polar coordinates of the electron with respect to the nucleus (Figure 15.1). Except in the case of a circular orbit, the

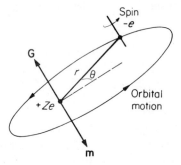

Figure 15.1. Schematic representation of the orbital motion of an electron about the nucleus and the spin of the electron about its own axis

angular velocity $\dot\theta$ is not constant. The area A enclosed by the orbit is given by

$$A = \int_0^{2\pi} \tfrac{1}{2} r^2 \, d\theta$$

$$= \int_0^{\tau} \tfrac{1}{2} r^2 \frac{d\theta}{dt} \, dt$$

$$= \frac{G}{2m} \int_0^{\tau} dt$$

$$= \frac{G\tau}{2m}$$

Magnetic Materials

By Ampere's law, a current I in a loop circuit of area A produces a magnetic field which, at some distance from the loop, is identical with that due to a magnetic dipole of moment IA. Therefore, the electron in its orbit has a magnetic dipole moment

$$\mathbf{m} = -\frac{e\mathbf{G}}{2m}$$

which is in a direction opposed to $\mathbf{G}$ as in Figure 15.1.

The ratio of the two moments, the *magnetogyric ratio*, is

$$\gamma = \frac{\mathbf{m}}{\mathbf{G}}$$

$$= -\frac{e}{2m}$$

The angular momentum is measured in units of $h/2\pi$ (see Section 2.4) so that a convenient unit which has been adopted for magnetic dipole moments is the *Bohr magneton* β, where

$$\beta = \frac{eh}{4\pi m}$$

$$= 9{\cdot}273 \times 10^{-24} \text{ A m}^2$$

An electron also has spin, with angular momentum $\tfrac{1}{2}h/2\pi$ (see Section 4.3), which, on the Bohr model, aligns itself either parallel or antiparallel to the orbital motion as in Figure 15.2. The magnetogyric ratio for spin is twice that for orbital motion, so that the corresponding magnetic moment is $\mp\beta$.*

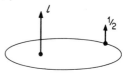

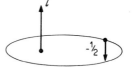

Figure 15.2. Parallel and antiparallel orientations of spin to orbital motion

---

* A nucleus also possesses spin, which is a simple multiple of $h/2\pi$, and has an associated magnetic moment. The magnitude of one nuclear magneton is

$$\frac{eh}{4\pi m_p} = 5{\cdot}05 \times 10^{-27} \text{ A m}^2$$

and, being much smaller than the Bohr magneton, its effect can usually be neglected.

In an atom with several electrons, the orbital angular momenta are restricted to certain relative directions (corresponding to the different permitted values of the m_l quantum number). The orbital angular momenta may then be combined to give a single resultant $Lh/2\pi$ (see Appendix 10). Similarly, the spin angular momenta may be combined to give a single resultant $Sh/2\pi$.

It is important to note here also that for any one electron shell, states with identical spins are filled first (Hunt's rule). Thus, for example, in chromium all five 3d electrons have spins of the same sign and in iron two of the 3d electrons are paired and the other four have spins of the same sign. The electrons arrange themselves to give the maximum possible total spin angular momentum consistent with Pauli's exclusion principle.

Also there can be coupling between the orbital and spin angular momenta to give a single resultant $Jh/2\pi$ (see Appendix 10).

In the case of atoms or ions which have any occupied s, p, d and f levels completely filled, then there is spherical distribution of charge and pairing of spins so that both L and S are zero and the magnetic moment is zero. Nevertheless, as will be seen in the next section, such materials have a susceptibility due to induced magnetism.

15.3 Diamagnetism

When the electron orbit with magnetic moment **m** is in a magnetic flux **B** such that the angular momentum vector **G** makes an angle θ to the flux direction (Figure 15.3), then there will be a couple **m** × **B** acting on the orbit which will

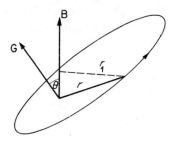

Figure 15.3. Vectors giving Larmor precession

produce a gyroscopic precession of the orbit around the direction of the **B** vector with an angular velocity

$$\omega = -\frac{m}{G}B$$

$$= \left(\frac{e}{2m}\right)B$$

Magnetic Materials

On account of this *Larmor precession*, the electron has an extra angular momentum, the instantaneous value of which is $mr_1^2\omega$ where r_1 is the projected length of the radius vector **r** on to a plane perpendicular to the field direction.

This extra angular momentum will also cause an additional magnetic moment

$$-\frac{e}{2m}mr_1^2\omega = \frac{e^2 r_1^2}{4m}\mathbf{B}$$

Now, r_1 is, in general, a variable so that the mean value of r_1^2 must be considered. The induced dipole moment is

$$\mathbf{m}_{ind} = -\frac{e^2}{4m}\overline{r_1^2}\mathbf{B}$$

The negative sign indicates that the induced moment is opposed to **B**, in agreement with Lenz's law.

In the case where there is spherical symmetry of charge, then summing over all the electrons in an atom, it can be shown that

$$\sum \overline{r_1^2} = \tfrac{2}{3}\sum \overline{r^2}$$

Hence the induced magnetic moment for the atom becomes

$$-\frac{e^2}{4m}\mathbf{B}\sum \overline{r_1^2} = -\frac{e^2}{6m}\mathbf{B}\sum \overline{r^2}$$

The total magnetic moment per unit volume is obtained by multiplying the induced magnetism per atom or molecule by N the number of atoms or molecules in unit volume, so that the *diamagnetic susceptibility* is given by

$$\chi_D = \frac{N\mathbf{m}}{\mathbf{H}}$$

$$= -\mu_0 N \frac{e^2}{6m}\sum \overline{r^2}$$

since $B/\mu_0 H \approx 1$ for a diamagnetic material.

While all materials will have this diamagnetic susceptibility, the only ones in which it is the only contribution to the total susceptibility are those which have complete electron shells or complete pairing of electrons in the atoms or molecules, i.e. inert gases, ionic compounds such as alkali halides in which all the ions have an inert gas structure and compounds such as organic materials, the molecules of which have entirely covalent bonding.

Since χ_D is proportional to $\overline{r^2}$, the outer electrons make the greatest contribution to the susceptibility. Also because $\sum \overline{r^2}$ varies very little with temperature, diamagnetic susceptibility is independent of temperature.

If a piece of material of volume V is placed in an inhomogeneous field, the force F_x which acts on the material in the x direction is

$$F_x = \chi_D V \frac{B}{\mu_0} \frac{dB}{dx}$$

For a diamagnetic material, χ_D is negative, so that F will act in the direction of decreasing B, i.e. the piece of material is repelled by the field.

Superconducting materials have a diamagnetic susceptibility equal to -1, i.e. the induced magnetism is sufficient to oppose completely the external flux density so that there is zero resultant flux in the material.

15.4 Paramagnetism

A whole atom or molecule which has a permanent dipole moment **m** would tend to rotate so that the dipole lines up parallel to the field direction, this alignment being opposed by thermal agitation. Materials containing such atoms or molecules would have a positive susceptibility, a phenomenon known as *paramagnetism*.

When treated in a classical manner and assuming that there is no appreciable interaction between the magnetic moments of adjacent atoms or molecules, the arguments are exactly the same as those discussed in Section 12.4 for the polarization of electric dipoles in an electric field. Except for very high field intensities or low temperatures, the susceptibility is

$$\chi_P = \frac{N\mu_0 \mathbf{m}^2}{3kT} \qquad (15.1)$$

A wave-mechanical treatment allowing for the quantization of the interaction between dipoles and the field (Appendix 11) gives a similar relationship. χ_P is of a much higher order of magnitude than the diamagnetic susceptibilities as may be seen by considering material composed of atoms each with a single electron orbit of moment 1 Bohr magneton and of radius equal to the radius a of the ground-state orbit of the hydrogen atom (see Section 3.5). Then the susceptibilities of such a material are

$$\chi_P = 8 \cdot 70 \times 10^{-33} N \quad \text{at} \quad 27\,°C$$

and

$$\chi_D = 1 \cdot 66 \times 10^{-35} N$$

where N is the number of atoms per m^3. Hence the paramagnetic susceptibility at room temperature is of the order of 500 times the diamagnetic susceptibility and so the former effect, when present, predominates.

Magnetic Materials

It will be seen that the paramagnetic susceptibility as given above varies inversely with absolute temperature, in agreement with the experimentally determined Curie law,

$$\chi_P = \frac{C}{T} \qquad (15.2)$$

published in 1895, which applies to paramagnetic gases and dilute solutions containing magnetic atoms or ions. In these cases, the interatomic distances are large so that the magnetic interaction between atoms can be assumed to be negligible.

In solids, however, the effective magnetic field which acts on a dipole includes the effect of the magnetization of the rest of the solid. Weiss showed (Appendix 12) that this would give a modified form of the Curie law with temperature dependence of susceptibility given by the *Curie–Weiss law*

$$\chi_P = \frac{C}{T - \theta}$$

This law holds experimentally only for values of T which are considerably higher than θ. For most paramagnetic materials, θ is very low and may even be negative.

15.5 Paramagnetism of simple metals

In the preceding section, the atoms or molecules with permanent dipole moments were considered and shown to be paramagnetic. In the case of a metal crystal, the valence electrons are not attached to particular atoms and so will orient themselves to the applied magnetic field independently. If ions have fully occupied electron levels and so zero magnetic moments (see Section 15.2), then the paramagnetic susceptibility will be due entirely to the electrons in the conduction band which are free to adopt either sign of spin, each thereby contributing a magnetic moment of $\pm \beta$.

In the absence of an external field, there will be equal numbers of spins of each sign and the density of states of the electrons of each spin will be some function of energy, which is shown as parabolic in Figure 15.4(a), all states up to the Fermi level E_0 being occupied when $T = 0$ K. When an external field is applied, each spin will adopt a parallel or antiparallel orientation with respect to the field and the potential energy of each electron will change by $\pm \beta B$ (Figure 15.4(b)). There will be a redistribution of electrons between the two spin states as shown in Figure 15.4(c) giving an excess of states of one sign.

Now βB is small compared with the spread of energy levels in the band (even for a very high flux density of $B = 10$ T, βB is only about 6×10^{-4} eV)

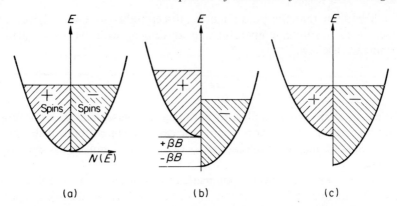

Figure 15.4. Density of + and − electron spins: (**a**) in the absence of an external field, (**b**) immediately on application of an external field, (**c**) after attainment of equilibrium

and at temperatures above 0 K there will be a Fermi distribution of the occupancy of the higher energy states which affects only a small fraction of the total number of electrons (see Section 10.5). Hence this paramagnetic susceptibility will be relatively independent of temperature and small compared with that discussed previously.

This was found to be so for the alkali metals and some of the alkaline earth metals. However, the noble metals, copper, silver and gold, are diamagnetic because the diamagnetic contribution due to the filled d shells is sufficiently large to outweigh the paramagnetic term.

The transition metals and rare earths have incomplete electron shells and can form paramagnetic compounds. The d and f levels are concentrated in a narrow energy band near the Fermi level, so that a change of B affects a larger proportion of the electrons. The paramagnetism of the rare earths is generally greater because there can be up to 14 electrons in an f level but only 10 in a d level.

When a specimen of paramagnetic material is placed in an inhomogeneous magnetic field, it will experience a force which attracts it in the direction of increasing field strength.

15.6 Ferromagnetism

For the rare earth metals just considered, where the paramagnetism is due to unpaired $4f$ electrons, the outer electrons in the $n = 5$ and $n = 6$ shells give sufficient screening for the interaction between the dipoles of adjacent atoms to be of no consequence except at very low temperatures. In certain of the transition metals, however, the unpaired $3d$ electrons are not so effectively

Magnetic Materials

screened and there is an appreciable interaction between neighbouring atoms, so that the unpaired spins are oriented nearly parallel to one another and tend to point in the same direction, giving domains which exhibit spontaneous magnetization, even in the absence of an externally applied field. Materials exhibiting such behaviour, of which iron is the principal one, are known as *ferromagnetic*.

The spontaneous magnetization exists only below a certain temperature known as the Curie temperature. As the external field is changed the magnetization of a ferromagnetic material changes in a unique manner, the susceptibility not being constant. The form of variation is discussed later.

A phenomenological explanation of ferromagnetism was given in 1907 by Weiss, who introduced the concept of an internal magnetic field which provided the alignment force on the dipoles. Although at that time the origin of the *Weiss molecular field* was not understood, the model has successfully explained many of the observed phenomena, such as spontaneous magnetization (Appendix 13) and the Curie–Weiss law, which was given in Section 15.4. The constant θ is the Curie temperature. For values of T greater than θ, ferromagnetic materials are paramagnetic with a susceptibility which varies according to the Curie–Weiss law and is independent of **B**.

If M is the magnetization of a domain at a temperature $T(<\theta)$ and M_s is the saturation magnetization at $T = 0$ when thermal agitation is absent and all dipoles are aligned, then the variation of M/M_s as a function of T/θ is approximately the same for iron, cobalt and nickel and an analysis of this variation suggests that electron spin makes the only significant contribution to the magnetism (see Appendix 13).

15.7 Exchange forces

The quantum-mechanical interaction between electron spins was treated by Heisenberg in 1926 who showed that the interaction energy depends critically upon the extent of the overlap of the wave functions, i.e. it depends upon the interatomic distance and may change sign as this distance is varied. As two atoms approach each other, the energy is decreased if spins of unpaired electrons assume a parallel orientation, but at closer distances an antiparallel orientation is favoured.

In a metal, the wave functions for all the outer electrons of one atom, including the $3d$ electrons of the transition metals, overlap sufficiently with those of neighbouring atoms for the electrons to behave as quasi-free electrons, obeying Fermi–Dirac statistics. Hence the energies must be considered in terms of band theory. If the exchange energy favours alignment of spins, then for there to be an excess of spins of one sign, one half band must contain more electrons than the other and hence be filled to a higher energy value, i.e. the decrease of energy due to interchange action has to be offset against this increase of energy. The existence or not of a net spin moment depends upon which of these two energies has the greater magnitude. The s and p energy bands are broad so that the kinetic energy involved in achieving a net spin would be greater than the decrease in exchange energy, but the d bands are sufficiently narrow for the spin polarization to give a net decrease in energy. This is shown diagrammatically in Figure 15.5.

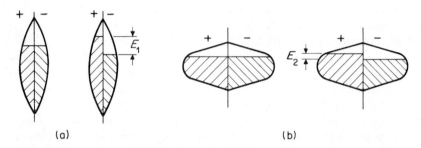

Figure 15.5. Schematic diagram of (**a**) wide, and (**b**) narrow energy bands. In each, n electrons have been removed from the negative to the positive spin half band to give a net spin moment. $E_1/2n$ is approximately the added kinetic energy in (**a**) and is larger than the corresponding energy $E_2/2n$ in (**b**)

The extent to which the d band is broadened is a function of the ratio of the interatomic spacing to the radius of the $3d$ orbit in the atom. The net energy of magnetization is also a function of this ratio and varies in the manner shown qualitatively in Figure 15.6.

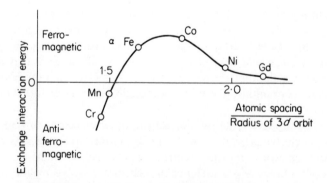

Figure 15.6. Variation of interaction energy with ratio atomic spacing/radius of $3d$ orbit for the iron series of transition metals

Magnetic Materials

Conditions favour parallel orientation of spins when the ratio is greater than 1·5. Hence iron, cobalt and nickel are ferromagnetic. For larger values of the ratio, the interaction is so small that any materials would be paramagnetic. Manganese is not ferromagnetic although compounds of it in which the atomic separation is increased may be—examples are Cu_2MnSn and Cu_2MnAl. In iron oxide, FeO, the spacing of the iron atoms is such that the material is ferromagnetic.

Gadolinium is ferromagnetic below 16 °C and some of the other rare earths are ferromagnetic at very low temperatures.

Iron, cobalt and nickel have six, seven and eight $3d$ electrons, respectively, and each has two $4s$ electrons. As the interaction favours parallel spins, one would expect there to be four, three and two unpaired $3d$ electrons, respectively, so that the atomic moments should be 4, 3 and 2 Bohr magnetons, respectively.

The actual values are, however, 2·2, 1·7 and 0·6 units, which can be explained in terms of the zone theory. The $3d$ and $4s$ energy bands overlap, the $3d$ band being much narrower, as was shown in Figure 10.5. The two bands are occupied to an extent such that the Fermi level is the same in each. Of the eight electrons which are available to these two states in iron, there are on average 7·4 in the $3d$ band and 0·6 in the $4s$ band. The spins of the electrons in the $4s$ band are paired, but of those in the $3d$ band there are 4·8 with spins of one sign and 2·6 with spins of the other sign, giving the net spin moment of 2·2 Bohr magnetons per atom.

Fractional numbers of electrons in the states implies that the electron distribution in the solid is given by a linear combination of $3d$ and $4s$ wave functions such that an electron spends part of its time in the $3d$ layer and part in the $4s$ layer.

15.8 Anisotropy energy

In single crystals of ferromagnetic material, the magnetic properties are anisotropic, there being *easy* and *hard* directions of magnetization, as may be seen from the curves in Figure 15.7.

Iron (b.c.c.) is most easily magnetized along the [100] direction and is most difficult along the [111] direction, while the reverse applies in nickel (f.c.c.). The easy direction in hexagonal close-packed cobalt is [001] and the magnetization in a perpendicular direction is very difficult.

Because of this magnetic anisotropy, the spontaneous magnetization will always tend to align with an easy direction.

The anisotropy arises because the d orbitals will conform to certain crystallographic directions due to the crystalline fields. Then the spin–orbital coupling causes the spin magnetic moments to exhibit anisotropy. As the external field is increased, the coupling is gradually overcome so that eventually the saturation

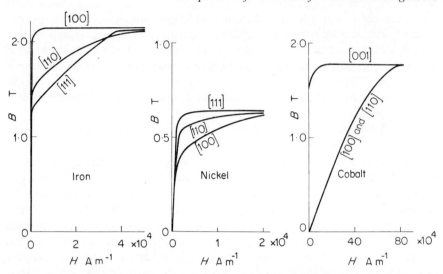

Figure 15.7. Magnetization curves for single crystals of iron, nickel and cobalt in the principal crystallographic directions at 20 °C

magnetization is the same in all directions. The direction of easy magnetization is not necessarily that of least atomic spacing. Thus for iron, the most difficult direction is [111] which is the one of least spacing.

The energy required to magnetize a specimen will therefore differ according to the direction of magnetization, the difference in energy being known as the *anisotropy energy*.

15.9 Magnetostriction

When a ferromagnetic crystal is magnetized, the physical dimensions change due to the interaction forces of the dipoles. The variation of this *magnetostriction* with magnetization due to the field acting along various crystallographic directions is shown for iron and nickel in Figure 15.8. The strain along the direction of magnetization is accompanied by strain of opposite sign in perpendicular directions so that the volume is unchanged. It is found that ferromagnetic alloys which have an integral number of Bohr magnetons per atom exhibit the smallest magnetostriction.

Magnetostriction causes a change in the effective anisotropy energy, the direction of deformation being such that this energy is reduced. The total change of energy due to magnetostriction is known as the *magnetostrictive energy*.

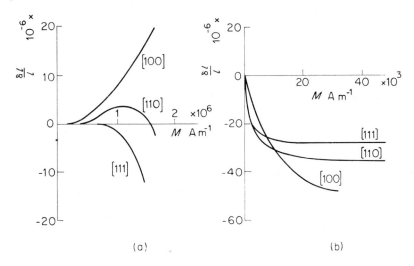

Figure 15.8. Magnetostriction of (a) iron, and (b) nickel single crystals. Ordinate gives strain in direction of applied field

15.10 Domains

For a material which can show spontaneous magnetization, a condition of zero magnetization would be unstable; nevertheless, specimens of ferromagnetic materials of large dimensions may have zero magnetization in the absence of an external magnetic field. This paradox was resolved by Weiss who postulated that such a specimen is composed of a number of small regions, called *domains*, each of which is magnetized to saturation, but that these are oriented in such a manner that the vector sum of the individual magnetic moments is zero.

The domain structure is that for which the total magnetic energy is a minimum. This magnetic energy includes the anisotropy and magnetostrictive energies already considered as well as two others.

The *magnetostatic energy* is the energy associated with the magnetization that appears at each end of the domains. In the case of a block of material which is a single domain, it would be the energy of the field in the surrounding space. If the block has several domains such that the magnetic flux follows closed paths within the material with no external flux, then the magnetostatic energy is zero.

Domains are separated by intermediate regions known as *domain walls* in which there is a gradual transition of the orientation of magnetization. Because adjacent dipoles are not parallel, then there is extra *exchange energy* involved in a domain wall. The energy per unit area of wall is minimized by making a more gradual change of angle between adjacent dipoles, i.e. it varies inversely as the thickness of the wall. However, in the domain wall, some dipoles will not be in easy directions of magnetization, so that extra anisotropy energy is involved. The optimum wall thickness will be that for which the sum of these two energies is a minimum. For iron this thickness is of the order of 1000 Å and the energy is about 10^{-3} J m^{-3}.

Domain structures tend to favour patterns such as those shown in Figure 15.9 with closure domains at the ends so that the magnetostatic energy is reduced to zero. The closure domains introduce extra energy and if the direction of magnetization in the closure domains is not the easy direction, anisotropy energy is introduced. Also the magnetostriction of the closure domains is not compatible with that of the other domains, causing internal stresses and an associated magnetostrictive energy. The domain size is determined by a balance between these energies, which are proportional to the closure domain volume and the wall energy. Narrower domains as in Figure 15.9(b) reduce the volume of closure domains but increase the domain wall area.

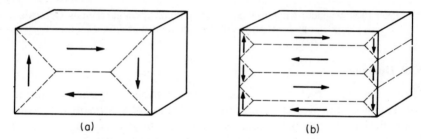

Figure 15.9. Possible domain structures in a single crystal

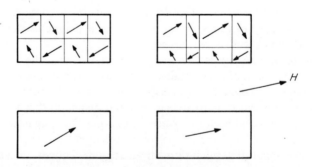

Figure 15.10. Stages in magnetization of ferromagnetic material, showing migration of domain walls followed by rotation of magnetization direction

Magnetic Materials

When an external magnetic field is applied, the domain walls migrate in such a manner that the domains with a direction of magnetization parallel or nearly parallel to the applied field grow at the expense of the remainder, the extent of migration at each stage being such as to minimize the total energy. When the domain walls have migrated as far as possible or vanished, the magnetization direction rotates until it becomes parallel to the applied field, when saturation is reached. This sequence is illustrated in Figure 15.10.

15.11 Ferromagnetic materials

In polycrystalline materials, the grains will generally have random crystallographic orientations, so that some have easy and some have hard directions of magnetization aligned with any external field direction. Also any magnetization will be complicated by the necessity of compatability of strain due to the magnetostriction of adjacent grains and of energy minimization, and by the movement of domain walls past lattice defects and non-magnetic inclusions. All these factors add up to the typical magnetization behaviour shown in Figure 15.11, where the flux density $B \, (= \mu_0(H + M))$ is shown as a function of

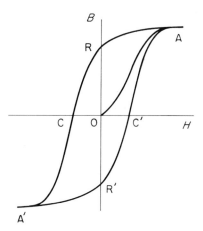

Figure 15.11. Typical hysteresis loop of ferromagnetic material

the magnetizing field H. For an initially unmagnetized specimen, as H is increased from zero, B increases corresponding to the movement of domain walls, at first with an increasing rate and then with a decreasing rate corresponding to domain rotation and the approach to saturation. When H is decreased to zero, then, due to the various effects listed above, the specimen does not return to a minimum energy system of domains with no external magnetic moment. Instead there is a residual magnetism, the *remnance*, and the material behaves as a *permanent magnet*. To reduce the magnetism to zero (point C) it is necessary to apply a reverse magnetizing force, H_c, which is known as the *coercive force*. Further increase of H in a negative direction causes a rapid increase of reverse magnetization, finally reaching saturation at A' if a sufficiently large field is applied. When H is again changed to a similarly large positive value, the similarly shaped path $A'R'C'A$ is followed.

This phenomenon is known as *magnetic hysteresis* and the closed curve is the *hysteresis loop*. When a specimen is taken through a complete cycle, there is an energy loss which is equal to

$$W = \oint dW$$

$$= V \oint H \, dB$$

= (Volume of specimen) × (Area of hysteresis loop)

If the domain walls are easily moved then the coercive force is low and the material is easy to magnetize, whereas if the domain walls are difficult to move then the coercive force is high and the material is more difficult to magnetize. Such materials are said to be magnetically soft and hard, respectively. Any structural defects, such as dislocations, non-magnetic inclusions and precipitates of a non-magnetic phase, contribute to the immobilization of domain boundaries. If a domain boundary passes through a non-magnetic inclusion, the inter-domain boundary area is reduced and so the boundary energy is also

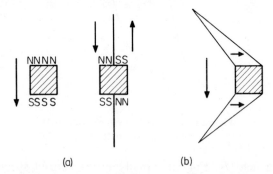

Figure 15.12. (a) Attraction of domain wall to inclusion reduces magnetostatic energy by causing more favourable distribution of north and south poles, (b) removal of poles by formation of spike-shaped domains. Magnetostatic energy reduced at expense of domain boundary energy

Magnetic Materials

reduced. To move the boundary away from the inclusion would require extra energy.

At the surface of the inclusion there would be free magnetization appearing and a finite magnetostatic energy. If the domain boundary passes through the inclusion, then there is a redistribution of magnetization which reduces the magnetostatic energy (see Figure 15.12(a)). Also it is found that spike-like domains with a different direction of magnetization form at inclusions and these reduce the magnetostatic energy at the expense of extra domain boundary energy (see Figure 15.12(b)). A passing domain boundary would be attracted to such a spike and require a finite critical energy to move away from it. The factors which make materials magnetically hard are also those which generally increase the mechanical hardness.

15.12 Application of ferromagnetic materials

For optimum performance in any particular application, a magnetic material with the most suitable magnetization curve should be chosen. Some examples are considered in the following paragraphs.

For a permanent magnet we require a magnetically hard material. As well as the obvious property of large remnance, it must also have a large coercive force to increase the difficulty of demagnetization due to any magnetic fields as occur in motors, magnetic chucks, etc. Such materials have large B–H loops.

Low alloy steels containing 0·6–1·0 per cent. carbon which are hardened by quenching or by precipitation hardening are the most widely used, but some of the best permanent magnet materials are of the Alnico and Alcomax types which contain aluminium, cobalt, nickel and copper in addition to iron. By suitable heat treatment in the presence of an external magnetic field, a precipitate forms which divides the structure into needle-like or plate-like fine particles, the behaviour of which is discussed in the next paragraph.

If the material is composed of very fine particles which are too small to have domain walls within them, then magnetization can take place only by rotation of the atomic magnets. Because this rotation must be through a direction of hard magnetization, the coercive force must be high and for small particles has been calculated to be about 500 times that for a large iron specimen. With suitably shaped particles in the form of thin rods oriented in the direction of magnetization, this coercive force could be increased still further.

Because most permanent magnet materials are mechanically hard and brittle, they must be cast to shape and finished by grinding. Alternatively they can be made by powder metallurgy methods from particles having the correct small size to give high coercive forces.

For materials which are to be used as cores in transformers and electrical machinery, the hysteresis and eddy current losses must be small—both for economies of power consumption and because lost energy is converted to heat, i.e. magnetically soft materials are needed. Silicon iron is superior to ordinary iron in this respect because

    a. the resistivity is greater, causing less eddy current loss,
    b. the hysteresis loop is smaller and
    c. the permeability is greater.

Further benefits can be gained by reducing the impurity content and also by orienting the grains by cold rolling so that the easy direction [100] is along the magnetization axis. The optimum silicon content is about 3 per cent.

The iron–nickel alloys (permalloys) have higher initial permeabilities than the iron–silicon alloys, the peak value being at about 79 per cent. nickel. These alloys have a face-centred crystal structure and magnetic properties analogous to nickel so that the directions of easy magnetization and minimum magnetostriction (see Figures 15.7 and 15.8) coincide which still further makes domain boundary migration easy. When the alloys are quenched so that no ordering of the iron and nickel atoms can take place, better results are obtained because

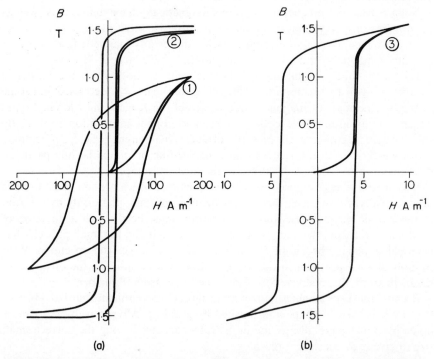

Figure 15.13. Typical hysteresis loops for various magnetic materials: (**a**) magnetically soft materials (1) pure iron, (2) 'H.C.R.' cold-rolled 50 per cent. Ni alloy; (**b**) magnetically hard material (3) Alcomax

Magnetic Materials

then the magnetic anisotropy and magnetostriction are smaller still. By annealing the material in a magnetic field, the maximum permeability can be increased many times over and an almost square hysteresis loop obtained.

Typical hysteresis loops for various ferromagnetic materials are shown in Figure 15.13.

15.13 Exchange mechanisms

From Figure 15.6 it is seen that for substances such as manganese and chromium the spins of adjacent atoms would be oriented in an antiparallel manner, whereas in iron, cobalt and nickel, the exchange energy is such that parallel orientations are favoured. Manganese and chromium would thus have a zero magnetic moment if the temperature is sufficiently low (below 40 °C for Cr and 100 K for Mn). Such materials are termed *antiferromagnetic*.

At interatomic distances larger than those shown on Figure 15.6, the direct interactions would be negligible, but there are many compounds, such as metal oxides, chlorides and sulphides, in which magnetic ions are separated by non-magnetic anions. These anions act as intermediaries and play a part in *superexchange interaction*.

Consider an oxide MO of a metal with an incomplete d shell which has a lattice similar to that of common salt (Figure 7.20(a)). Each oxygen ion is surrounded by six metal ions in an octahedral configuration. The metal ions are too far apart for their $3d$ orbitals to overlap, but they will overlap the orbitals of the neighbouring oxygen ions.

Each oxygen ion will have ten electrons in all, two each in the $1s$ and $2s$ levels and six in the $2p$ level. Because of the influence of the crystalline field, the p orbitals will be directed along the three coordinate axes of the lattice so that each of the pairs of p orbitals will overlap the d orbitals of two metal ions as shown diagrammatically in Figure 15.14.

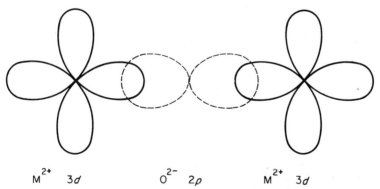

Figure 15.14. Wave function of one pair of $2p$ electrons of oxygen ion overlapping wave functions of $3d$ electrons of metal ions

There is probably a transfer of a *p* electron from the oxygen to a *d* level in one of the metal ions which leaves the oxygen with a net magnetic moment so that there will be exchange forces between it and the other metal ion. The result, which has been explained at least qualitatively on a wave-mechanical basis, is that the two metal ions adopt an antiparallel alignment.

This M—O—M coupling is stronger than that of the oblique M—M coupling and dictates the spin pattern. The metal ions may then be considered as two sub-lattices of ions each having all the ions of one spin (Figure 15.15).

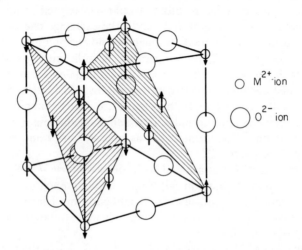

Figure 15.15. Cubic lattice of MO. Two sub-lattices with opposite spin orientations are indicated by shaded planes

In a material like this, where each sub-lattice has the same magnetic moment (of which manganous oxide is an example), the material will have no net spontaneous magnetization and is antiferromagnetic.

15.14 Susceptibility of antiferromagnetic material

It is found experimentally that at higher temperatures, where thermal agitation can overcome the ordering of the spins, the material is paramagnetic with a relationship of the Curie–Weiss type

$$\chi = \frac{C}{T - \theta}$$

but θ is a negative quantity. This relationship is followed down to the Néel temperature which is defined as that below which both sub-lattices have spontaneous magnetization.

Magnetic Materials

Below the Néel temperature, the susceptibility is a function of the direction of the applied magnetic field. There will be preferred directions for the spontaneous magnetization of the antiferromagnetic sub-lattices, just as there are preferred directions for ferromagnetism. If the applied field is perpendicular to the magnetization direction, it tends to rotate the dipoles of both sub-lattices and will be almost independent of temperature. When the applied field is parallel to the magnetization, then at $T = 0$ K, the susceptibility will be zero because the only change can be to rotate the antiparallel dipoles through an angle of 180°, which would require an infinite force. As the temperature is increased, the dipoles are displaced slightly by thermal agitation and the susceptibility increases. For a polycrystalline sample, the susceptibility will be an average value for all orientations, which Van Vleck has shown to be

$$\chi = \tfrac{2}{3}\chi_\perp + \tfrac{1}{3}\chi_\parallel$$

where $\chi_\perp$ and $\chi_\parallel$ are the susceptibilities in directions which are perpendicular and parallel, respectively, to the magnetization direction. A typical curve for the variation of susceptibility with temperature is shown in Figure 15.16.

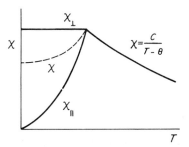

Figure 15.16. Typical temperature variation of susceptibility of an antiferromagnetic material

15.15 Ferrimagnetism

When a material has two sub-lattices with magnetizations which are antiparallel but unequal in intensity, then there is not complete compensation of the magnetic moments. Materials with this property are known as *ferrimagnetic*.

The susceptibility is a more complicated function of temperature than is the case for antiferromagnetism, being

$$\frac{1}{\chi} = \frac{T}{C} + \frac{1}{\chi_0} - \frac{K}{T - \theta}$$

where C, χ_0, K and θ are material constants. This represents a hyperbolic curve as shown in Figure 15.17.

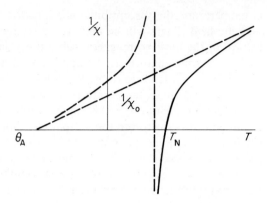

Figure 15.17. Temperature variation of the inverse susceptibility of a ferrimagnetic material. The full-line curve is similar to the observed behaviour in the paramagnetic region

At high temperatures, the material is paramagnetic, the susceptibility increasing as the temperature falls and becoming infinite at the Néel temperature T_N. At temperatures below T_N the material is spontaneously magnetized, the dipoles within each sub-lattice adopting parallel alignment.

15.16 Ferrimagnetic materials

Various magnetic oxides, which include ions of the iron group of transition metals, are ferrimagnetic. The ones of greatest technological importance to date are the *ferrites*, which are oxides with the spinel structure, the *rare earth–iron garnets* and those with a *magnetoplumbite* structure.

15.16.1 *Ferrites*

These compounds have a crystal structure which is isomorphic with spinel, $MgAl_2O_4$, and can be formed by substituting other divalent ions for magnesium and other trivalent ions for aluminium.

The unit cell, which is cubic, contains thirty-two oxygen anions, sixteen trivalent cations and eight divalent cations (Figure 15.18). The sites which are occupied by cations are of two types—tetrahedral sites surrounded by four equidistant oxygen ions and octahedral sites surrounded by six oxygen ions. These are referred to as A and B sites, respectively.

In the normal spinel structure, the divalent ions occupy the A sites and the trivalent ions occupy the B sites. However, the magnetic ferrites have the *inverted spinel structure* in which eight of the sixteen trivalent ions occupy all the A sites and the other eight together with the eight divalent ions occupy the B sites in random order.

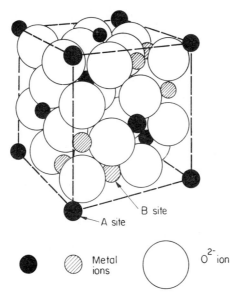

Figure 15.18. The unit cell of the spinel structure

In ferrites, the spins in the two lattices are antiparallel, all the spins within either one lattice being parallel. Then the magnetic moments of the ions in sub-lattice A cancel those of the trivalent ions in sub-lattice B and the net magnetic moment is that of the divalent ions in the B sites.

Thus magnetite, Fe_3O_4, consists of divalent and trivalent iron ions according to $Fe^{2+}O \cdot Fe^{3+}{}_2O_3$. Now Fe^{2+} and Fe^{3+} ions have six and five d electrons, respectively, and because in isolated atoms or ions with partly filled d shells the spins are always oriented to give the maximum number of unpaired spins, the magnetic moments are 4 and 5 Bohr magnetons, respectively. Hence the total magnetic moment of Fe_3O_4 would be 14 if the spins were all aligned. The experimental value of the magnetic moment is 4·2 which is almost equal to that of the Fe^{2+} ions alone.

By substitution of the di- or trivalent iron ions of magnetite with others, ferrites with a wide range of magnetic properties can be produced.

Examples are manganese zinc ferrite and nickel zinc ferrite which are magnetically soft, the former having a permeability in the range 650–2300 and the latter a permeability of about 100. The value of the permeability depends strongly on the magnetostriction, smaller magnetostriction causing a greater μ_r. Thus magnetite has positive magnetostriction, but in manganese zinc ferrite it is negative. Increasing the Fe_2O_3 content of the latter material reduces the magnetostriction and increases the permeability. The nickel zinc ferrite has a larger magnetostriction still further increased by increasing the nickel or zinc content.

Some ferrites such as manganese magnesium ferrite and manganese copper ferrite are magnetically hard.

15.16.2 Magnetic garnets

The magnetic garnets are oxides having crystal structures isomorphous with garnet $Ca_3Fe_2(SiO_4)_3$ and having ferrimagnetic properties which are in some respects superior to those of the ferrites. The general formula is $M_3Fe_2(FeO_4)_3$ where M is a trivalent ion of yttrium or any rare earth metal. Yttrium and lutetium are non-magnetic so that in garnets containing these elements the only magnetic ions are Fe^{3+} which form two antiparallel sub-lattices and give magnetization curves of the type shown in Figure 15.19(a). The other rare earth ions are magnetic and when present form a third sub-lattice which (except in the case of europium–iron garnets) causes the resulting curve to be of the type shown in Figure 15.19(b). The garnets are transparent.

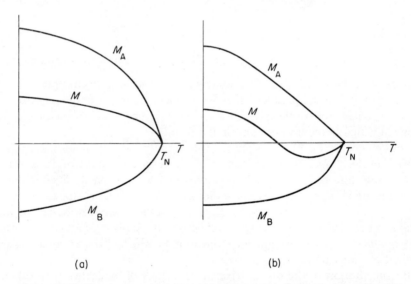

Figure 15.19. Temperature variation of magnetization of some ferrites below the Néel temperature: (**a**) normal type, (**b**) anomalous type with reversal of sign of M at compensation point. M_A and M_B are the respective magnetizations of the A and B sub-lattices

15.16.3 Magnetoplumbites

These are ferrimagnetic oxides with hexagonal crystal structures similar to magnetoplumbite $PbFe_{7.5}Mn_{3.5}Al_{0.5}Ti_{0.5}O_{19}$. The best known is $BaFe_{12}O_{19}$

Magnetic Materials

which, having a high coercive force, is a permanent magnet material. Its trade name is Ferroxdure. This material has a very large magnetic anisotropy, the easy direction of magnetization being parallel to the hexagonal axis. By applying a high static magnetic field during shaping operations the easy direction of magnetization can be put in any desired direction. The raw materials are readily available and so the material is cheap. Addition of zinc or magnesium causes the direction of easy magnetization to be changed.

15.17 Spin resonance

The Larmor precession of an electron due to its orbital motion has been described in Section 15.3. The precession due to the total angular motion will have an angular velocity

$$\omega = \gamma B$$

where γ is the magnetogyric ratio given by

$$\gamma = \frac{ge}{2m}$$

The Landé splitting factor g takes account of the spin–orbital coupling (see Appendix 10).

If an alternating magnetic field is applied in a direction at right-angles to B with an angular frequency equal to the Larmor frequency, then the amplitude of precession will grow and energy will be absorbed from the oscillating field. In a solid, due to the interaction of the dipole moments of adjacent atoms, there will be damping effects and a limiting amplitude will be reached.

When the frequency of the alternating field equals that of the spins, then there will be resonance and maximum absorption of energy. At frequencies near to this resonant frequency, the susceptibility has a complex form similar to the complex electrical susceptibility discussed in Section 12.5. Studies of resonant frequencies yield much information about the spin process.

15.18 Electromagnetic wave propagation in ferrites

When an alternating magnetic field is applied to a ferrite, the extra energy given to the dipoles is in part transferred to nearby dipoles so that a magnetic wave travels along the solid. The manner of propagation is influenced by any externally applied static magnetic field.

Any plane wave can be considered as the sum of plane polarized waves of which there are three types according to the direction of propagation and the plane of polarization relative to the direction of the polarizing field H. These are

a. waves travelling parallel to H, for which the direction of the magnetic vector is always perpendicular to H

b. waves travelling perpendicular to H for which the plane of polarization may be either parallel or perpendicular to H.

15.18.1 *Propagation parallel to polarizing field*

It can be shown that circularly polarized electromagnetic waves will propagate along the direction of the static polarizing field. For waves polarized in the same sense as the Larmor precession the effective permeability is greater and hence the velocity of propagation is less than for waves polarized in the opposite sense.

Now any plane-polarized wave can be considered as the sum of two circularly polarized waves of opposite sign which will traverse the material with different velocities. As, however, the frequency remains constant, the two components have a relative phase shift which varies along the path, so that the plane of polarization of the resultant wave is rotated from the original direction (Figure 15.20). This is known as *Faraday rotation* and occurs for all paramagnetic

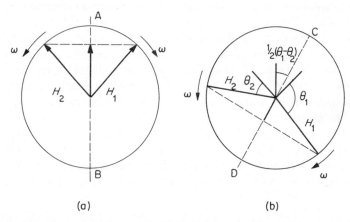

Figure 15.20. (a) Two circularly polarized waves represented by vectors H_1 and H_2 rotating in opposite directions at equal speeds combine to form a plane-polarized wave. (b) After travelling a distance in the ferrite, the vectors differ in phase by θ_1 and θ_2, respectively, from (a) and the plane of polarization is rotated by $\frac{1}{2}(\theta_1 - \theta_2)$

and ferromagnetic materials, but not for diamagnetic materials. The effect is usually small and, provided that the frequency is not close to that of resonance, the absorption is very slight.

Magnetic Materials

15.18.2 *Propagation perpendicular to polarizing field*

Of the two components mentioned earlier, that with its magnetic vector perpendicular to the polarizing field has a propagation velocity which depends on the magnitude of H, while that with its magnetic vector parallel to H is unaffected by the magnitude of H. Hence for unpolarized waves travelling in this direction there is a birefringence effect, the incident wave being split into two components which are refracted differently.

15.19 Applications of ferrites

Disadvantages of the use of ferromagnetic materials in transformers, etc., are the losses due to hysteresis and eddy currents. The former can be reduced by choosing materials with small hysteresis loops and the latter by using laminations and by increasing the electrical resistance, e.g. by using silicon iron. As eddy current loss is proportional to (frequency)^2, it becomes much more significant at radio frequencies. The specific resistance of ferrites is about 10^{16} times that of metals at room temperature so that eddy currents are almost negligible even at radio frequencies. Hence the magnetically soft ferrites can be used as transformer cores, etc., at such high frequencies. Nickel zinc ferrites with high nickel and zinc contents can be used at frequencies up to 1000 MHz. In ferrites, there can be an exchange of electrons between Fe^{2+} and Fe^{3+} ions, giving a slight conductivity, an effect which is absent from the garnets where all the iron ions are trivalent. Hence the latter are more useful at microwave frequencies. Also the permeabilities are high compared with iron dust cores so that smaller volumes can be used for transformers with the same inductance, aiding miniaturizing of components.

Many ferrites have *B–H* curves which are almost rectangular in shape. When materials with these 'square' loops are magnetized in one direction, there is no change when a reverse field is applied unless this exceeds the coercive force, when the magnetization becomes saturated in the reverse direction. Because of the absence of eddy currents, this change can be extremely rapid, taking 1 μs or less. Ferrite toroids are used in digital computers and automatic telephone exchanges as memory cores for storing information. The direction of magnetization can be either $+$ or $-$, these two states representing alternatives, such as 0 and 1 in a binary digital system.

Ferrites find many applications in microwave devices. The Faraday rotation has already been mentioned. If a rod of suitable size is placed in a waveguide and a plane-polarized wave is incident upon it, then the plane of polarization will be rotated through an angle determined by the material and by the static magnetic field applied in the direction of propagation. This is a non-reciprocal effect. If a rotation of θ takes place for a wave travelling in one direction, then if the wave is reflected, a further rotation of θ will take place so that the initial and final waves have planes of polarization inclined at 2θ to one another.

This principle can be used in a *circulator*—a device containing a number of ports with the behaviour that an input to the nth port gives an output from the $(n + 1)$th port and zero output from any other.

Cavities containing ferrites will have resonant frequencies which can be varied by changing the polarizing field. Hence they can be used to tune cavities and in tunable filters.

When the r.f. magnetizing field is sufficiently large, the response of the ferrites is non-linear and this property can be used for various purposes which require non-linearity, e.g. frequency doublers, detectors, amplifiers and power limiters.

The materials that are used for their non-reciprocal properties should have as high a saturation magnetization as possible, a property found in mixed ferrites such as nickel–zinc and manganese–magnesium ferrites. The latter have lower dielectric loss and are preferred for low-power applications, but have low Curie points and cannot be used for high-power applications.

Garnets are more suitable than ferrites for non-linear applications as they will operate at a lower level of power.

Questions for Chapter 15

1. On the assumption that each electron of a helium atom occupies a $1s$ state as though the other electron were not present, calculate the susceptibility of helium gas at a pressure of 10^5 N m^{-2} and a temperature of 27 °C.

2. For a gas, each molecule of which has a magnetic moment of 2 Bohr magnetons, calculate the volume susceptibility of the gas at standard temperature and pressure.

3. For a free nickel atom in the normal state the S, L and J values are 1, 3 and 4, respectively. Find the Landé splitting factor.

4. From the saturation flux densities of iron, cobalt and nickel as given in Figure 15.7 determine for each element the number of Bohr magnetons per atom.

5. Determine the anisotropy energy per unit volume for magnetization in [001] and [100] directions of cobalt due to a field of 800×10^3 A m^{-1}.

6. Show that the following values of magnetic susceptibility for nickel are consistent with the Curie law. Find the Curie constant and temperature

T	(K)	800	900	1000	1100	1200
χ_m		$10^{-5} \times$ 3·3	2·1	1·55	1·2	1·0

Appendix One

SI Units

The units used in this book are, with a few exceptions, units of the *Système International d'Unites*. There are six basic units, from which all the others are derived and which are:

>Length—metre (m)
>Mass—kilogramme (kg)
>Time—second (s)
>Thermodynamic temperature—kelvin (K)
>Electrical current—ampere (A)
>Luminous intensity—candela (cd)

These are defined as follows:

metre—1 650 763·73 wavelengths *in vacuo* of the radiation corresponding to the transition between the energy levels $2p_{10}$ and $5d_5$ of the krypton-86 atom.

kilogramme—the mass of the international prototype which is in the custody of the *Bureau International des Poids et Mesures* (BIPM) at Sèvres, France.

second—the duration of 9 192 631 770 cycles of the radiation corresponding to the transition between the two hyperfine levels of the fundamental state of the caesium-133 atom.

kelvin—the fraction 1/273·16 of the thermodynamic temperature of the triple point of water.

ampere—the constant current which, if maintained in two straight parallel conductors of infinite length, of negligible circular cross-section, and placed at a distance of 1 m apart in vacuum, would produce between these conductors a force equal to 2×10^7 N m^{-1}.

candela—the luminous intensity, in the perpendicular direction, of a surface of 1/600 000 m^2 of a black body at the freezing temperature of platinum under a pressure of 101 325 N m^{-2}.

The derived units which appear in this book are:

hertz (Hz)—unit of frequency. The number of repetitions of a regular occurrence in one second.

newton (N)—unit of force. That force which applied to a mass of 1 kg gives it an acceleration of 1 m s^{-2}.

joule (J)—unit of energy. The work done when the point of application of a force of 1 N is displaced by 1 m in the direction of the force.

watt (W)—unit of power, equal to 1 J s^{-1}.

coulomb (C)—unit of electric charge. The quantity of electricity transported in 1 s by 1 A.

volt (V)—unit of electric potential. The potential difference between two points of a conducting wire carrying a constant current of 1 A when the power dissipated between these points is 1 W.

ohm (Ω)—unit of electric resistance. The resistance between two points of a conductor when a constant potential difference of 1 V between the two points produces in this conductor a current of 1 A.

farad (F)—unit of electrical capacitance. The capacitance of a capacitor between the plates of which there appears a potential difference of 1 V when it is charged by 1 coulomb of electricity.

weber (Wb)—unit of magnetic flux. The magnetic flux which, linking a circuit of one turn, produces in the circuit an electromotive force of 1 V as it is reduced to zero at a uniform rate in 1 s.

henry (H)—unit of electrical inductance. The inductance of a closed circuit in which an electromotive force of 1 V is produced when the electric current in the circuit varies uniformly at 1 A s^{-1}.

tesla (T)—unit of magnetic flux density, equal to 1 Wb m^{-2}.

Other units which are used in the book are:

ångström (Å)—10^{-10} m.

litre (l)—10^{-3} m^3.

electron-volt (eV)—the energy acquired by an electron in traversing a potential difference of 1 V ($= 1.602 \times 10^{-19}$ J).

degree Celsius (°C)—(K − 273·16).

kilogramme mole (kmol)—a mass of substance which expressed in kg is numerically equal to the molecular weight of the substance.

rydberg $= R_H hc = 2.18 \times 10^{-18} \text{ J} = 13.6 \text{ eV}$.

Appendix Two

Fundamental Constants and Conversion Constants

Electronic charge	e	$1\cdot60202 \times 10^{-19}$ C
Rest mass of electron	m	$9\cdot1083 \times 10^{-31}$ kg
Rest mass of proton	m_p	$1\cdot67238 \times 10^{-27}$ kg
Rest mass of neutron	m_n	$1\cdot67470 \times 10^{-27}$ kg
Avogadro's number	N_0	$6\cdot0225 \times 10^{26}$ kmol^{-1}
Planck's constant	h	$6\cdot625 \times 10^{-34}$ J s
Boltzmann's constant	k	$1\cdot38048 \times 10^{-23}$ J K^{-1}
Stefan–Boltzmann constant	σ	$5\cdot67 \times 10^{-8}$ W m^{-2} K^{-4}
Velocity of light	c	$2\cdot99793 \times 10^{8}$ m s^{-1}
Permittivity of space	ε_0	$8\cdot854 \times 10^{-12}$ F m^{-1}
Permeability of space	μ_0	$4\pi \times 10^{-7}$ H m^{-1}
Unified atomic mass unit	m_u	$1\cdot6605 \times 10^{-27}$ kg
Electron-volt		$1\cdot60202 \times 10^{-19}$ J
Gas constant	$\overline{R}$	8314 J K^{-1} kmol^{-1}
Rydberg constant	R_∞	$1\cdot09736 \times 10^{-7}$ m^{-1}
Bohr magneton	β	$9\cdot273 \times 10^{-24}$ A m^2

APPENDIX THREE

Solution of Schrödinger's Equation for the Hydrogen Atom

The wave functions required are the solutions of equation (3.14), with the potential given by equation (3.13), the boundary conditions given in equation (3.15) and the normalization condition given in equation (3.16).

It is assumed that the variables are separable and that the wave function can be expressed as the product of $R(r)$, $\Theta(\theta)$ and $\Phi(\phi)$, which are functions of r, θ and ϕ, respectively.

Equation (3.14) then becomes

$$\frac{1}{r^2}\frac{d}{dr}\left(r^2\frac{dR}{dr}\right)\Theta\Phi + \frac{1}{r^2\sin\theta}\frac{d}{d\theta}\left(\sin\theta\frac{d\Theta}{d\theta}\right)R\Phi$$

$$+ \frac{1}{r^2\sin^2\theta}\frac{d^2\Phi}{d\phi^2}R\Theta + \frac{8\pi^2 m}{h^2}(E-V)R\Theta\Phi = 0$$

By rearranging the terms and multiplying by $r^2\sin^2\theta/R\Theta\Phi$, the equation is changed to

$$\sin^2\theta\left\{\frac{1}{R}\frac{d}{dr}\left(r^2\frac{dR}{dr}\right) + \frac{1}{\Theta\sin\theta}\frac{d}{d\theta}\left(\sin\theta\frac{d\Theta}{d\theta}\right) + r^2\frac{8\pi^2 m}{h^2}(E-V)\right\}$$

$$= -\frac{1}{\Phi}\frac{d^2\Phi}{d\phi^2} \quad (A.1)$$

The left-hand side is a function of r and θ only while the right-hand side is a function of ϕ only. They can be equal for all values of r, θ and ϕ only if each side equals a constant. If this constant is taken to be m_l^2, then the equation for Φ is

$$\frac{d^2\Phi}{d\phi^2} + m_l^2\Phi = 0$$

the solution of which is

$$\Phi = A \cos m_l\phi + B \sin m_l\phi$$

Applying the boundary condition given in equation (3.15) that Φ is a single-valued function of ϕ, i.e. its value is the same when ϕ is changed by 2π, the only possible solutions are those for which $m_l = 0$ or a positive or negative integer. We will take the solution in the form

$$\Phi = A\, e^{im_l\phi}$$

By putting the left-hand side of equation (A.1) equal to m_l^2 also and again separating variables

$$\frac{1}{R}\frac{d}{dr}\left(r^2\frac{dR}{dr}\right) + r^2\frac{8\pi^2 m}{h^2}(E - V) = \frac{m_l^2}{\sin^2\theta} - \frac{1}{\Theta\sin\theta}\cdot\frac{d}{d\theta}\left(\sin\theta\frac{d\Theta}{d\theta}\right) \quad \text{(A.2)}$$

Here, the left-hand side is a function of r only and the right-hand side is a function of θ only. Again each side must equal a constant, which it is convenient to take as $l(l + 1)$. The equation for Θ is then

$$\frac{1}{\sin\theta}\frac{d}{d\theta}\left(\sin\theta\frac{d\Theta}{d\theta}\right) + \left\{l(l+1) - \frac{m_l^2}{\sin^2\theta}\right\}\Theta = 0$$

The solution of this type of differential equation is known as a Legendre polynomial $P_l^{m_l}(\cos\theta)$. The only solutions which are acceptable in this case are those for which the value of the function does not become infinite within the relevant range of θ, which is 0 to π. This is so only when l is integral and $l \geqslant m_l$. For small values of l and m_l, these functions are fairly simple, thus:

$$P_0^0(\cos\theta) = 1$$
$$P_1^0(\cos\theta) = \cos\theta$$
$$P_1^1(\cos\theta) = \sin\theta$$
$$P_2^0(\cos\theta) = \tfrac{1}{2}(3\cos^2\theta - 1)$$
$$P_2^1(\cos\theta) = 3\sin\theta\cos\theta$$
$$P_2^2(\cos\theta) = 3\sin^2\theta$$

If $l = 1$, then m_l can be either 0, 1 or -1 so that the function $\Theta\Phi$ can have either of the three forms

$$\cos\theta \qquad \sin\theta\, e^{i\phi} \qquad \sin\theta\, e^{-i\phi}$$

If the wave functions ψ_1 and ψ_2 are two solutions for the same energy E, i.e. they are degenerate, then any linear combination of ψ_1 and ψ_2 is also a solution of the equation for the same value of E. It is often more convenient to replace the complex functions given for $m_l = \pm 1$ by real linear combinations of them. The $\Theta\Phi$ functions for $m_l = \pm 1$ then become

$$\tfrac{1}{2}\sin\theta(e^{i\phi} + e^{-i\phi}) = \sin\theta\cos\phi$$

and

$$\tfrac{1}{2}i\sin\theta(e^{i\phi} - e^{-i\phi}) = \sin\theta\sin\phi$$

Each of these corresponds to a nodal plane passing through the origin. Thus $\cos\theta = 0$ when $\theta = \pi/2$, i.e. the nodal plane is perpendicular to the z axis. The other two solutions will have nodal planes which pass through the z axis and which are perpendicular to the x and y axes, respectively (see Figure A.1).

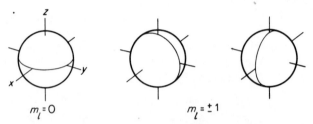

Figure A.1. Nodal planes for p wave functions. $m_l = 0$, $\Theta\Phi = \cos\theta$: $m_l = \pm 1$, $\Theta\Phi = \sin\theta\cos\phi$ and $\sin\theta\sin\phi$

For $l = 2$, five angle functions are possible, corresponding to $m_l = 0, \pm 1, \pm 2$, the forms of the solutions and the corresponding nodal planes being shown in Figure A.2.

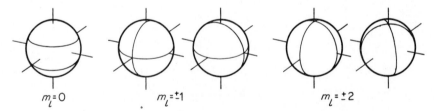

Figure A.2. Nodal planes for d wave functions. $m_l = 0$, $\Theta\Phi = \tfrac{1}{2}(3\cos^2\theta - 1)$: $m_l = \pm 1$, $\Theta\Phi = 3\sin\theta\cos\theta\cos\phi$ and $3\sin\theta\cos\theta\sin\phi$; $m_l = \pm 2$, $\Theta\Phi = 3\sin^2\theta\cos 2\phi$ and $3\sin^2\theta\sin 2\phi$

In general, for one value of l, there are $(2l + 1)$ possible functions of θ and ϕ. By equating the left-hand side of equation (A.2) to $l(l + 1)$, we have

$$\frac{1}{r^2}\frac{d}{dr}\left(r^2\frac{dR}{dr}\right) + \left\{\frac{8\pi^2 m}{h^2}(E - V) - \frac{l(l + 1)}{r^2}\right\}R = 0 \qquad (A.3)$$

Solution of Schrodinger's Equation for the Hydrogen Atom

Substitution of $R(r) = f(r)/r$ in equation (A.3) gives

$$\frac{d^2f}{dr^2} + \left\{ \frac{8\pi^2 m}{h^2}(E - V) - \frac{l(l+1)}{r^2} \right\} f = 0 \qquad (A.4)$$

a simpler equation which is similar to the one-dimensional potential well considered in Section 3.3, with effective potential energy given by

$$V' = \left\{ V + \frac{h^2}{8\pi^2 m} \cdot \frac{l(l+1)}{r^2} \right\}$$

The effective potential wells for different values of l will take the forms shown in Figure A.3. The acceptable solutions will be those which decay to zero as $r \to \infty$ and which are finite for $r \to 0$.

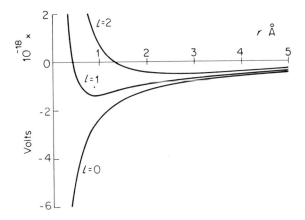

Figure A.3.

If equation (A.4) is rewritten in terms of a new variable x, where

$$x = \frac{2\pi}{h}\sqrt{(2m|E|)}\,r \qquad (A.5)$$

and the value of V inserted from equation (3.13), it is considerably simplified to

$$\frac{d^2f}{dx^2} + \left\{ -1 + \frac{\alpha}{x} - \frac{l(l+1)}{x^2} \right\} f = 0 \qquad (A.6)$$

where $|E|$ has been substituted for E and

$$\alpha = \frac{e^2 2\pi}{4\pi\varepsilon_0 h}\sqrt{(2m/|E|)}$$

$$= \frac{e^2}{2\varepsilon_0 h}\sqrt{(2m/|E|)} \qquad (A.7)$$

When x is large, the terms in $1/x$ and $1/x^2$ can be neglected compared with one so that the equation becomes

$$\frac{d^2 f}{dx^2} - f = 0$$

the solutions of which are

$$f = A\,e^{\pm x}$$

The required solution that satisfies the boundary condition that $f \to 0$ as $x \to \infty$ is therefore

$$f = A\,e^{-x}$$

Hence we can put

$$f = e^{-x} g(x)$$

which, substituted in equation (A.6), gives

$$\frac{d^2 g}{dx^2} - 2\frac{dg}{dx} + \left\{\frac{\alpha}{x} - \frac{l(l+1)}{x^2}\right\} g = 0 \qquad (A.8)$$

If we try to solve g as a polynomial in x

$$g = x^\sigma(a_0 + a_1 x + a_2 x^2 + \ldots a_n x^n + \ldots)$$

Substituting for g in equation (A.8) and equating coefficients of $x^{\sigma-2}$ and of $x^{\sigma-2+s}$ gives

$$\sigma(\sigma - 1) = l(l+1) \qquad (A.9)$$

and

$$a_s[(\sigma + s)(\sigma + s - 1) - l(l+1)] = a_{s-1}[2(\sigma + s - 1) - \alpha] \qquad (A.10)$$

From equation (A.9),

$$\sigma = l + 1 \quad \text{or} \quad -l$$

Now the required solution must approach 0 as $r \to 0$ so that R can be finite at $r = 0$. Hence σ must be positive and

$$\sigma = l + 1$$

Solution of Schrödinger's Equation for the Hydrogen Atom

Then equation (A.10) becomes

$$a_s[(s + l + 1)(s + l) - l(l + 1)] = a_{s-1}[2(s + 1) - \alpha] \quad (A.11)$$

If $2(s + l) - \alpha$ does not become zero for any integral value of s, then when s is large

$$\frac{a_s}{a_s - 1} \simeq \frac{2}{s}$$

The ratio of successive terms is the same as the ratio of the higher order successive terms in the expansion of e^{2x}. Thus g tends to infinity in the same manner as e^{2x}. Since $f = e^{-x}g$, then f and hence ψ will increase in the manner of e^x and hence tend to infinity as $x \to 0$, which is not an acceptable solution.

If, however, $2(s + l) - \alpha$ vanishes for any integral value of s, then the series will terminate and the function $f(x)$ will tend to zero as $x \to \infty$. The condition is that

$$\alpha = 2(s + l)$$

where $s = 1, 2, 3$, etc.

Then from equation (A.7)

$$E = -\frac{me^4}{8\varepsilon_0^2 h^2} \cdot \frac{1}{(s + l)^2} \quad (A.12)$$

The energy thus depends upon a single number n defined by

$$n = s + l$$

If $n = 1$, then $l = 0$ and $\alpha = 1$. From equation (A.11)

$$a_1 = a_0 \left(\frac{2 - 2}{2 - 0}\right) = 0$$

and all higher coefficients are also zero. Hence

$$f = e^{-x}a_0 x$$

Now from equations (A.5) and (A.12)

$$x = \frac{\pi me^2}{\varepsilon_0 h^2 n} r = r/an$$

where

$$a = \frac{\varepsilon_0 h^2}{\pi me^2} = 0.53 \text{ Å}$$

and in this case $n = 1$.

Hence
$$R = A e^{-r/a}$$
where A is an arbitrary constant. The wave function is thus
$$\psi = A e^{-r/a}$$
Normalization of a function expressed in spherical polar coordinates becomes
$$\int_{r=0}^{\infty} \int_{\theta=0}^{\pi} \int_{\phi=0}^{2\pi} |\psi|^2 |r^2 \sin\theta \, dr \, d\theta \, d\phi = 1$$
which, for $n = 1$, is
$$\int_0^\infty \int_0^\pi \int_0^{2\pi} A^2 e^{-2r/a} r^2 \sin\theta \, dr \, d\theta \, d\phi$$
$$= 2\pi A^2 \int_0^\infty \int_0^\pi e^{-2r/a} r^2 \sin\theta \, dr \, d\theta$$
$$= 4\pi A^2 \int_0^\infty e^{-2r/a} r^2 \, dr$$
$$= 4\pi A^2 \left[-\frac{a}{2} e^{-2r/a} r^2 + a \int e^{-2r/a} r \, dr \right]_0^\infty$$
$$= 4\pi A^2 \left[-\frac{a}{2} e^{-2r/a} r^2 + a \left\{ -\frac{a}{2} e^{-2r/a} r + \frac{a}{2} e^{-2r/a} \, dr \right\} \right]_0^\infty$$
$$= 4\pi A^2 \left[-\frac{a}{2} e^{-2r/a} r^2 - \frac{a^2}{2} e^{-2r/a} r - \frac{a^3}{4} e^{-2r/a} \right]_0^\infty$$
$$= \pi a^3 A^2 = 1$$
Therefore
$$\psi = (1/\pi a^3)^{1/2} e^{-r/a}$$
is the wave function corresponding to $n = 0, l = 0$.

APPENDIX FOUR

Solution of Schrödinger's Equation for Simple Harmonic Oscillator

Schrödinger's equation is

$$\frac{d^2\psi}{dx^2} + \frac{8\pi^2 m}{h^2}(E - \tfrac{1}{2}\alpha x^2)\psi = 0$$

and the boundary conditions are that $\psi = 0$ for $x = \pm\infty$. By making the substitutions

$$A = 2\left(\frac{4\pi^2 m}{\alpha h^2}\right)^{1/2} E \qquad (A.13)$$

and

$$y = \left(\frac{4\pi^2 m\alpha}{h^2}\right)^{1/4} x$$

Schrödinger's equation is simplified to

$$\frac{d^2\psi}{dy^2} + (A - y^2)\psi = 0 \qquad (A.14)$$

At large values of y, the energy E can be neglected when compared with the potential energy and the equation is

$$\frac{d^2\psi}{dy^2} = y^2 \psi$$

The solution of this which satisfies the boundary condition is

$$\psi = C\,e^{-(1/2)y^2} \qquad (A.15)$$

The exact function is determined by replacing the constant C in equation (A.15) by a suitable function $C(y)$ of y which, substituted in equation (A.14), gives

$$\frac{d^2\psi}{dy^2} + (A - y^2)\psi = e^{-(1/2)y^2}\left[C(A-1) - 2y\frac{dC}{dy} + \frac{d^2C}{dy^2}\right] = 0$$

The expressions within the square brackets, when equated to zero, is the differential equation for $C(y)$, i.e.

$$\frac{d^2C}{dy^2} - 2y\frac{dC}{dy} + (A-1)C = 0 \tag{A.16}$$

$C(y)$ may be expressed as a polynomial in y

$$C = \sum_n a_n y^n$$

and when substituted in the left-hand side of equation (A.16) gives another polynomial in y which can be zero only if the coefficient of each power of y is zero. Thus for the coefficient of y^n

$$(n+2)(n+1)a_{n+2} = [(2n+1) - A]a_n$$

If the solution for ψ is to converge to 0 at $y = \infty$, then the polynomial $C(y)$ must have a finite number of terms. If n_1 is the maximum value of n, then $a_{n_1+2} = 0$, which will be so if

$$A = 2n_1 + 1 \tag{A.17}$$

Again if n_2 is the smallest value of n, then $a_{n_2-2} = 0$ or

$$n_2(n_2 - 1) = 0$$

i.e. $n_2 = 0$ or 1 so that the series contains positive powers of y only (a necessary condition as otherwise ψ would be infinity at $y = 0$).

Substituting from equation (A.13) into equation (A.17)

$$2\left(\frac{4\pi^2 m}{\alpha h^2}\right)^{1/2} E = 2n + 1$$

or

$$E = \frac{h}{2\pi}\sqrt{\frac{\alpha}{m}}(n + \tfrac{1}{2})$$

$$= h\nu(n + \tfrac{1}{2})$$

where ν is the frequency of the oscillation (see equation (3.17)).
It can be shown that the function C_n is given by

$$C_n(y) = (-1)^n e^{y^2} \frac{d^n}{dy^n} e^{-y^2}$$

Solution of Schrödinger's Equation for Simple Harmonic Oscillator

When expressed in terms of the variable x, the wave functions after normalization are

$$\psi = \left(\frac{\beta^{1/2}}{2^n n . \sqrt{\pi}}\right)^{1/2} e^{-(1/2)\beta x^2} C_n(\beta^{1/2} x)$$

where

$$\beta^2 = 4\pi^2 m\alpha/h^2$$

APPENDIX FIVE

Proof of Spherical Symmetry of Completely Full p and d Levels

For the p states, the wave functions are

$$m_l = 0 \qquad \psi = R(r) \cos \theta$$
$$m_l = \pm 1 \qquad \psi = R(r) \sin \theta \cos \phi$$

and

$$\psi = R(r) \sin \theta \cos \phi$$

where $R(r)$ is the radial function.

If all the states of a level are occupied by electrons (six in all allowing for spins) then the probability of finding an electron in a volume element dV is

$$|\psi|^2 \, dV = 2R^2(\cos^2 \theta + \sin^2 \theta \cos^2 \phi + \sin^2 \theta \sin^2 \phi) \, dV$$
$$= 2R^2 \, dV$$

which is independent of the angular coordinates. That is the electron-charge distribution is a function of r only and is spherically symmetrical.

For the d states, the wave functions are

$$m_l = 0 \qquad \psi = R(r)\tfrac{1}{2}(3\cos^2\theta - 1)$$

$$m_l = \pm 1 \qquad \psi = R(r)\sqrt{3}\sin\theta\cos\theta \begin{array}{l} \cos\phi \\ \sin\phi \end{array}$$

$$m_l = \pm 2 \qquad \psi = R(r)\frac{\sqrt{3}}{2}\sin^2\theta \begin{array}{l} \cos 2\phi \\ \sin 2\phi \end{array}$$

Proof of Spherical Symmetry for Completely Full p and d Levels

The electronic charge density at one point due to ten electrons in these states is

$$|\psi|^2 \, dV = 2R^2[\tfrac{1}{4}(3\cos^2\theta - 1)^2 + 3\sin^2\theta\cos^2\theta(\cos^2\phi + \sin^2\phi)$$
$$+ \tfrac{3}{4}\sin^4\theta(\cos^2 2\phi + \sin^2 2\phi)]$$
$$= 2R^2[\tfrac{9}{4}\cos^4\theta - \tfrac{3}{2}\cos^2\theta + \tfrac{1}{4} + 3\cos^2\theta(1 - \cos^2\theta)$$
$$+ \tfrac{3}{4}(1 - \cos^2\theta)^2]$$
$$= 2R^2$$

Hence the electron charge in a fully-occupied d level is also spherically symmetrical.

APPENDIX SIX

The Electron Structure of the Elements

Z	Element	1 s	2 s	2 p	3 s	3 p	3 d	4 s	4 p	4 d	4 f	5 s	5 p	5 d	5 f	6 s	6 p	6 d	6 f	7 s	7 p
1	H	1																			
2	He	2																			
3	Li	2	1																		
4	Be	2	2																		
5	B	2	2	1																	
6	C	2	2	2																	
7	N	2	2	3																	
8	O	2	2	4																	
9	F	2	2	5																	
10	Ne	2	2	6																	
11	Na	2	2	6	1																
12	Mg	2	2	6	2																
13	Al	2	2	6	2	1															
14	Si	2	2	6	2	2															
15	P	2	2	6	2	3															
16	S	2	2	6	2	4															
17	Cl	2	2	6	2	5															
18	A	2	2	6	2	6															
19	K	2	2	6	2	6		1													
20	Ca	2	2	6	2	6		2													
21	Sc	2	2	6	2	6	1	2													
22	Ti	2	2	6	2	6	2	2													
23	V	2	2	6	2	6	3	2													
24	Cr	2	2	6	2	6	5	1													
25	Mn	2	2	6	2	6	5	2													
26	Fe	2	2	6	2	6	6	2													

The Electron Structure of the Elements

Z	Element	1	2		3			4				5				6				7	
		s	s	p	s	p	d	s	p	d	f	s	p	d	f	s	p	d	f	s	p
27	Co	2	2	6	2	6	7	2													
28	Ni	2	2	6	2	6	8	2													
29	Cu	2	2	6	2	6	10	1													
30	Zn	2	2	6	2	6	10	2													
31	Ga	2	2	6	2	6	10	2	1												
32	Ge	2	2	6	2	6	10	2	2												
33	As	2	2	6	2	6	10	2	3												
34	Se	2	2	6	2	6	10	2	4												
35	Br	2	2	6	2	6	10	2	5												
36	Kr	2	2	6	2	6	10	2	6												
37	Rb	2	2	6	2	6	10	2	6			1									
38	Sr	2	2	6	2	6	10	2	6			2									
39	Y	2	2	6	2	6	10	2	6	1		2									
40	Zr	2	2	6	2	6	10	2	6	2		2									
41	Nb	2	2	6	2	6	10	2	6	4		1									
42	Mo	2	2	6	2	6	10	2	6	5		1									
43	Ma	2	2	6	2	6	10	2	6	6		1									
44	Ru	2	2	6	2	6	10	2	6	7		1									
45	Rh	2	2	6	2	6	10	2	6	8		1									
46	Pd	2	2	6	2	6	10	2	6	10											
47	Ag	2	2	6	2	6	10	2	6	10		1									
48	Cd	2	2	6	2	6	10	2	6	10		2									
49	In	2	2	6	2	6	10	2	6	10		2	1								
50	Sn	2	2	6	2	6	10	2	6	10		2	2								
51	Sb	2	2	6	2	6	10	2	6	10		2	3								
52	Te	2	2	6	2	6	10	2	6	10		2	4								
53	I	2	2	6	2	6	10	2	6	10		2	5								
54	Xe	2	2	6	2	6	10	2	6	10		2	6								
55	Cs	2	2	6	2	6	10	2	6	10		2	6			1					
56	Ba	2	2	6	2	6	10	2	6	10		2	6			2					
57	La	2	2	6	2	6	10	2	6	10		2	6	1		2					
58	Ce	2	2	6	2	6	10	2	6	10	2	2	6			2					
59	Pr	2	2	6	2	6	10	2	6	10	3	2	6			2					
60	Nd	2	2	6	2	6	10	2	6	10	4	2	6			2					
61	Pm	2	2	6	2	6	10	2	6	10	5	2	6			2					
62	Sm	2	2	6	2	6	10	2	6	10	6	2	6			2					
63	Eu	2	2	6	2	6	10	2	6	10	7	2	6			2					
64	Gd	2	2	6	2	6	10	2	6	10	7	2	6	1		2					
65	Tb	2	2	6	2	6	10	2	6	10	8	2	6	1		2					
66	Dy	2	2	6	2	6	10	2	6	10	10	2	6			2					
67	Ho	2	2	6	2	6	10	2	6	10	11	2	6			2					
68	Er	2	2	6	2	6	10	2	6	10	12	2	6			2					
69	Tm	2	2	6	2	6	10	2	6	10	13	2	6			2					

Z	Ele-ment	1	2		3			4				5				6				7	
		s	s	p	s	p	d	s	p	d	f	s	p	d	f	s	p	d	f	s	p
70	Yb	2	2	6	2	6	10	2	6	10	14	2	6			2					
71	Lu	2	2	6	2	6	10	2	6	10	14	2	6	1		2					
72	Hf	2	2	6	2	6	10	2	6	10	14	2	6	2		2					
73	Ta	2	2	6	2	6	10	2	6	10	14	2	6	3		2					
74	W	2	2	6	2	6	10	2	6	10	14	2	6	4		2					
75	Re	2	2	6	2	6	10	2	6	10	14	2	6	5		2					
76	Os	2	2	6	2	6	10	2	6	10	14	2	6	6		2					
77	Ir	2	2	6	2	6	10	2	6	10	14	2	6	7		2					
78	Pt	2	2	6	2	6	10	2	6	10	14	2	6	8		2					
79	Au	2	2	6	2	6	10	2	6	10	14	2	6	10		1					
80	Hg	2	2	6	2	6	10	2	6	10	14	2	6	10		2					
81	Tl	2	2	6	2	6	10	2	6	10	14	2	6	10		2	1				
82	Pb	2	2	6	2	6	10	2	6	10	14	2	6	10		2	2				
83	Bi	2	2	6	2	6	10	2	6	10	14	2	6	10		2	3				
84	Po	2	2	6	2	6	10	2	6	10	14	2	6	10		2	4				
85	At	2	2	6	2	6	10	2	6	10	14	2	6	10		2	5				
86	Rn	2	2	6	2	6	10	2	6	10	14	2	6	10		2	6				
87	Fr	2	2	6	2	6	10	2	6	10	14	2	6	10		2	6			1	
88	Ra	2	2	6	2	6	10	2	6	10	14	2	6	10		2	6			2	
89	Ac	2	2	6	2	6	10	2	6	10	14	2	6	10		2	6	1		2	
90	Th	2	2	6	2	6	10	2	6	10	14	2	6	10		2	6	2		2	
91	Pa	2	2	6	2	6	10	2	6	10	14	2	6	10	2	2	6	1		2	
92	U	2	2	6	2	6	10	2	6	10	14	2	6	10	3	2	6	1		2	
93	Np	2	2	6	2	6	10	2	6	10	14	2	6	10	5	2	6			2	
94	Pu	2	2	6	2	6	10	2	6	10	14	2	6	10	6	2	6			2	
95	Am	2	2	6	2	6	10	2	6	10	14	2	6	10	7	2	6			2	
96	Cm	2	2	6	2	6	10	2	6	10	14	2	6	10	7	2	6	1		2	
97	Bk	2	2	6	2	6	10	2	6	10	14	2	6	10	8	2	6	1		2	
98	Cf	2	2	6	2	6	10	2	6	10	14	2	6	10	10	2	6			2	
99	E	2	2	6	2	6	10	2	6	10	14	2	6	10	11	2	6			2	
100	Fm	2	2	6	2	6	10	2	6	10	14	2	6	10	12	2	6			2	
101	Mv	2	2	6	2	6	10	2	6	10	14	2	6	10	13	2	6			2	
102	No	2	2	6	2	6	10	2	6	10	14	2	6	10	14	2	6			2	

APPENDIX SEVEN

The Maxwell-Boltzmann Distribution

In a gas composed of one kind of molecule, the translational energy of a molecule of mass m is given by $\tfrac{1}{2}mc^2$ where c is its velocity. This energy changes at each collision but in a large population in thermal equilibrium, there is an equilibrium distribution of velocities.

From the Boltzmann distribution discussed in Appendix 8 (equation (A.22)) we see that the probability of there being a particle with velocity c is

$$p(c) = A\,e^{-Bc^2}$$

where A and B are constants. In a large population where there is a continuous range of velocities, we should be more precise and let $p(c)\,dc$ be the number of molecules with velocities in the range c to $c + dc$.

The velocity of a gas molecule is a vector quantity implying direction as well as speed. If the velocity vectors are represented in a three-dimensional velocity diagram as in Figure A.4, $p(c)\,dc$ is the number of molecules whose velocities

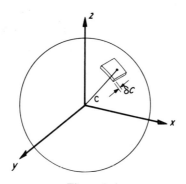

Figure A.4.

are represented by points lying in the element at the end of the vector **c**. This element is of unit area on the surface of the sphere of radius c, and has a thickness dc.

The total number of particles $N(c)\,dc$ with speeds in the range c to $c + dc$ is obtained by integrating all such elements lying between radii c and $c + dc$ over the whole of the sphere

$$N(c) = 4\pi c^2 p(c)$$
$$= 4\pi c^2 A\, e^{-Bc^2} \tag{A.18}$$

The constants A and B can be determined by calculating the total number of molecules and the total energy from equation (A.1). The kinetic energy of all the molecules is $\tfrac{3}{2}\bar{R}T$ per mole or $\tfrac{3}{2}NkT$ for N molecules, so that

$$N = \int_0^\infty N(c)\,dc \tag{A.19}$$

and

$$\tfrac{3}{2}NkT = \int_0^\infty \tfrac{1}{2}mc^2 N(c)\,dc \tag{A.20}$$

By substituting equation (A.18) into equations (A.19) and (A.20) and using the standard integrals

$$\int_0^\infty x^2 e^{-ax^2}\,dx = \frac{1}{4a}\left(\frac{\pi}{a}\right)^{1/2}$$

and

$$\int_0^\infty x^4 e^{-ax^2}\,dx = \frac{3}{8a^2}\left(\frac{\pi}{a}\right)^{1/2}$$

we get

$$B = m/2kT$$

and

$$A = N\left(\frac{m}{2\pi kT}\right)^{3/2}$$

so that

$$N(c) = 4\pi c^2 N\left(\frac{m}{2\pi kT}\right)^{3/2} e^{-mc^2/2kT}$$

APPENDIX EIGHT

Maxwell-Boltzmann and Fermi-Dirac Statistics

The distribution of the energies of the individual members of a collection of many particles is a problem in statistical mechanics. It depends upon the availability of permissible energy levels and also upon any restrictions placed upon the numbers which can occupy any particular level.

Consider first a system of particles for which there is no interaction except when they are in collision. This is so if the de Broglie wavelength is less than the average particle separation. Otherwise classical statistical mechanics cannot be used. Also let there be no restriction as to the numbers of particles which can occupy any particular energy level.

Consider two particles of energies E_1 and E_2 in collision. Energy will be conserved so that if δ is the change of energy of either particle and E_3 and E_4 are the respective energies after the collision, then

$$E_1 + E_2 = (E_1 + \delta) + (E_2 - \delta)$$
$$= E_3 + E_4$$

If $p(E)$ is the probability of there being a particle with energy E, then the number of such collisions in unit time is

$$Cp(E_1)p(E_2)$$

where C is a constant.

Now at equilibrium, the total number with each energy is constant and hence as many reverse collisions will occur. Then

$$p(E_1)p(E_2) = p(E_3)p(E_4)$$
$$= p(E_1 + \delta)p(E_2 - \delta) \tag{A.21}$$

This equation can be satisfied only by
$$p(E) = Ae^{\alpha E}$$
where A and α are constants. That this is a possible solution of equation (A.21) is seen by substitution, thus

$$\begin{aligned} p(E_1)p(E_2) &= A^2 \, e^{\alpha E_1} \, e^{\alpha E_2} \\ &= A^2 \, e^{\alpha(E_1+E_2)} \end{aligned}$$

and

$$\begin{aligned} p(E_3)p(E_4) &= A^2 \, e^{\alpha(E_1+\delta)} \, e^{\alpha(E_2-\delta)} \\ &= A^2 \, e^{\alpha(E_1+E_2)} \, e^{\alpha(\delta-\delta)} \\ &= A^2 \, e^{\alpha(E_1+E_2)} \end{aligned}$$

It can be shown (see Appendix 7) that
$$\alpha = -1/kT$$
so that
$$p(E) = Ae^{-E/kT} \tag{A.22}$$

is the probability that a particle will have an energy E. This is the Boltzmann distribution. It applies when considering any form of energy, kinetic, gravitational potential, electrical potential, etc. Also it applies whether the particles can have any energy value from a continuous spectrum or only certain discrete values.

Now if Pauli's exclusion principle holds, a collision of particles with energies E_1 and E_2 will result in a transfer of energy to give energies E_3 and E_4 only if the states of energy E_3 and E_4 are unoccupied.

Hence the number of collisions in unit time with this energy interchange is also dependent upon the probabilities of states E_3 and E_4 being vacant, and is
$$Cp(E_1)p(E_2)[1 - p(E_3)][1 - p(E_4)]$$

Again, equating this rate to the reverse rate gives
$$\left[\frac{1-p(E_3)}{p(E_3)}\right]\left[\frac{1-p(E_4)}{p(E_4)}\right] = \left[\frac{1-p(E_1)}{p(E_1)}\right]\left[\frac{1-p(E_2)}{p(E_2)}\right]$$

which can be satisfied only if
$$\frac{1-p(E)}{p(E)} = A\,e^{\beta E}$$
or
$$p(E) = \frac{1}{Ae^{\beta E} + 1}$$

Now if E is large, $A e^{\beta E} \gg 1$ and

$$p(E) \approx \frac{1}{A} e^{-\beta E}$$

which from the Maxwell–Boltzmann distribution gives

$$\beta = 1/kT$$

Also let

$$A = e^{-E_0/kT}$$

Then the probability becomes

$$p(E) = \frac{1}{e^{(E-E_0)/kT} + 1}$$

which is the Fermi–Dirac distribution function.
Now

$$1 - p(E) = 1 - \frac{1}{e^{(E-E_0)/kT} + 1}$$

$$= \frac{e^{(E-E_0)/kT} + 1 - 1}{e^{(E-E_0)/kT} + 1}$$

$$= \frac{1}{e^{-(E-E_0)/kT} + 1}$$

i.e. if

$$E_i - E_0 = E_0 - E_j$$

then

$$1 - p(E_j) = p(E_i)$$

so that the distribution is symmetrical about E_0 except for a change of sign.

APPENDIX NINE

Proof of Richardson–Dushmann Equation

In a metal, the number of electrons in unit volume with energies between E and $E + dE$ is given by the product of the density of states $N(E)$ (equation (10.1)), the Fermi–Dirac distribution function $p(E)$ (equation (9.1)) and dE, i.e.

$$N(E)p(E)\,dE = \frac{8\pi m\sqrt{2mE}}{h^3} \frac{dE}{e^{(E-E_0)/kT} + 1} \tag{A.23}$$

Now an electron with energy E has a speed s given by

$$E = \tfrac{1}{2}ms^2 \tag{A.24}$$

where the speed s is independent of the direction of the velocity and is given by

$$s^2 = v_x^2 + v_y^2 + v_z^2$$

v_x, v_y and v_z being the components of the velocity.

The number of electrons $N(s)\,ds$ in the speed range s to $s + ds$ is given by substituting equation (A.24) into equation (A.23)

$$N(s)\,ds = \frac{8\pi m^3}{h^3} \frac{s^2\,ds}{e^{(E-E_0)/kT} + 1} \tag{A.25}$$

The velocity of an electron can be represented as a point on a space velocity diagram, where the three coordinates of the point are proportional to the three components v_x, v_y and v_z of the velocity.

If the velocities of all the electrons are plotted on such a space velocity diagram (Figure A.5), then all the electrons with speeds between s and $s + ds$ will be represented by points which lie between the two concentric spherical shells of radii s and $s + ds$.

Proof of Richardson–Dushmann Equation

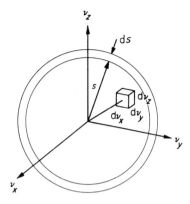

Figure A.5.

Also electrons which have velocities in the range v_x to $v_x + dv_x$, v_y to $v_y + dv_y$, v_z to $v_z + dv_z$ are represented on the same diagram by points which lie within the volume element dv_x, dv_y, dv_z.

Now the directions of motion are randomly distributed so that the ratio of the number within the velocity range **v** to **v** + d**v** to the number within the speed range s to $s + ds$ is the ratio of the volume of the element $dv_x\, dv_y\, dv_z$ to the volume of the spherical shell, where s is the speed equal to the magnitude of the velocity **v**. Hence the number $N(\mathbf{v})\, d\mathbf{v}$ in the range **v** to **v** + d**v** is given by

$$N(\mathbf{v})\, d\mathbf{v} = N(s)\, ds \frac{dv_x\, dv_y\, dv_z}{4\pi s^2\, ds}$$

Then from equation (A.25)

$$N(\mathbf{v})\, d\mathbf{v} = \frac{2m^3}{h^3} \frac{dv_x\, dv_y\, dv_z}{e^{(E-E_0)/kT} + 1} \qquad \text{(A.26)}$$

If an electron approaches a surface which is perpendicular to the x direction, then only the x component of the velocity is affected by the potential field at the surface. The electron can escape from the surface only if v_x is sufficiently large for

$$\tfrac{1}{2}mv_x^2 \geqslant eV_B$$

so that if v'_x is the lower limit of escape velocity, then

$$\tfrac{1}{2}mv'^2_x = eV_B$$

The number of electrons with an x component of velocity between v_x and $v_x + dv_x$ which strike unit area of the surface in unit time is equal to the number contained in a volume of dimensions v_x in the x direction and unit length in the y and z directions.

Then the total number which escape from unit area in unit time is given by integrating $v_x N(\mathbf{v})\, d\mathbf{v}$ between the limits of v'_x and infinity for v_x and over all values of v_y and v_z.

Also because V_B is sufficiently greater than E_0 for $E - E_0$ to be larger than kT in most cases, unity can be neglected compared with the exponential term in the denominator and the Fermi function can be approximated to $e^{-(E-E_0)/kT}$.

The current density J of the thermally emitted electrons is the product of the number escaping and the electronic charge, i.e.

$$J = e \int_{v'_x}^{\infty} \int_{-\infty}^{\infty} \int_{-\infty}^{\infty} v_x N(\mathbf{v})\, d\mathbf{v}$$

By substituting for $N(\mathbf{v})\, d\mathbf{v}$ from equation (A.26) and by using

$$E = \tfrac{1}{2} m(v_x^2 + v_y^2 + v_z^2)$$

the current density is

$$J = \frac{2em^3}{h^3} \int_{v'_x}^{\infty} \int_{-\infty}^{\infty} \int_{-\infty}^{\infty} v_x\, e^{-(mv_x^2 + mv_y^2 + mv_z^2 - 2E_0)/2kT}\, dv_x\, dv_y\, dv_z$$

$$= \frac{2em^3}{h^3} e^{E_0/kT} \int_{v'_x}^{\infty} v_x\, e^{-mv_x^2/2kT}\, dv_x \int_{-\infty}^{\infty} e^{-mv_y^2/2kT}\, dv_y$$

$$\int_{-\infty}^{\infty} e^{-mv_z^2/2kT}\, dv_z \qquad (A.27)$$

The integrals with respect to v_y and v_z are standard integrals, each with the value $(2\pi kT/m)^{1/2}$. The integral with respect to v_x becomes

$$-\frac{kT}{m}[e^{-mv_x^2/2kT}]_{v'_x}^{\infty} = \frac{kT}{m} e^{-mv_x'^2/2kT}$$

so that equation (A.27) becomes

$$J = \frac{2em^3}{h^3} e^{E_0/kT} \left(\frac{kT}{m} e^{-V_B/kT}\right)\left(\frac{2\pi kT}{m}\right)$$

which, because

$$E_W = V_B - E_0$$

becomes

$$J = \frac{4\pi e m k^2}{h^3} T^2\, e^{-E_W/kT}$$

or

$$J = A_0 T^2\, e^{-E_W/kT} \qquad (A.28)$$

Proof of Richardson–Dushmann Equation

A_0 is a universal constant with the value

$$A_0 = 1\cdot 2 \times 10^6 \text{ A m}^{-2} \text{ K}^{-2}$$

Equation (A.28) is the *Richardson–Dushmann* equation for thermionic emission.

APPENDIX TEN

Wave-Mechanical Treatment of Orbital and Spin Motions

Wave-mechanical theory gives the angular momentum of an electron with quantum number l as $l^*h/2\pi$ where

$$l^* = \sqrt{l(l+1)}$$

The direction of the angular momentum vector l^* is not defined except in the presence of an external magnetic field, when it can adopt any one of $2l+1$ orientations corresponding to possible values of the magnetic quantum number m_l, which can be any integer from $-l$ to $+l$. The component of the angular momentum in the direction of the field is $m_l h/2\pi$.

Thus for a p orbital ($l = 1$), the angular momentum is $\sqrt{2}h/2\pi$ and in the presence of a field can adopt any of three components $h/2\pi$, 0 or $-h/2\pi$ in the field direction (Figure A.6(a)). The interaction energy of the dipole with the field is different in the three cases and so the degenerate p level becomes split into three, causing a splitting of any spectral lines associated with electron jumps into or out of the orbit—the *Zeeman* effect.

The five possible components for a d orbital ($l = 2$) are shown in Figure A.6(b).

On a quantum-mechanical basis, the angular momentum due to spin is given by $s^*h/2\pi$, where

$$s^* = \sqrt{s(s+1)}$$

and s has the value $\tfrac{1}{2}$ (Figure A.6(c)).

If the *spin–orbital coupling* is so small that the orbital and spin moments react separately to an external magnetic field, then the spin motion will orient itself such that its component of angular momentum parallel to the field direction is $\pm m_s h/2\pi$, where the spin quantum number takes the values $m_s = \pm\tfrac{1}{2}$.

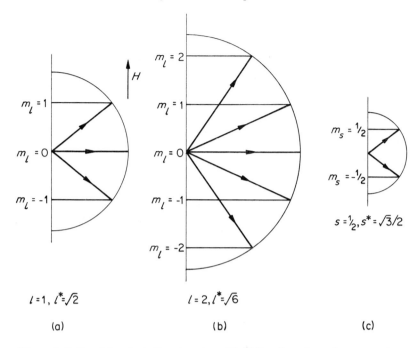

Figure A.6. Possible orientations to external field direction of angular momentum vectors: orbital angular momentum $l^*h/2\pi = \sqrt{l(l+1)}h/2\pi$ for **(a)** p orbital, and **(b)** d orbital; **(c)** spin moment $s^*h/2\pi = \sqrt{s(s+1)}h/2\pi$

On the other hand, where the spin–orbital coupling is strong, the angular momenta combine vectorially to give a single momentum vector **j***

$$\mathbf{j^*} = \mathbf{l^*} + \mathbf{s^*}$$

where

$$j^* = \sqrt{j(j+1)}$$

and j can take all half odd integral values from $l + s$ to $l - s$, i.e. for a single p state $j = 1\frac{1}{2}$ or $\frac{1}{2}$ so that $j^* = \sqrt{15/2}$ or $\sqrt{3/2}$. In the presence of a magnetic field, j^* can take orientations such that its components in the field direction are $m_j h/2\pi$ where m_j is a new quantum number taking all values from $+j$ to $-j$ in steps of unity.

The magnetogyric ratio γ for orbital motion is defined in Section 15.2. The ratio for spin is 2γ.

The magnetic moments due to orbital motion and spin will likewise combine vectorially as in Figure A.7. Then **l*** and **s*** will precess about the direction of **j***

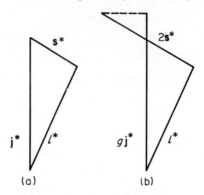

Figure A.7. Vectorial combination of (a) angular momenta and (b) magnetic moments due to orbital and spin motions. Values are in units of (a) $h/2\pi$, (b) β

so that the mean resultant of magnetic moment perpendicular to **j*** is zero, while that parallel to **j*** is $g\gamma\mathbf{j}^*$, where g is a constant known as the *Landé splitting factor*. It is easy to show that

$$g = 1 + \frac{j^{*2} + s^{*2} - l^{*2}}{2j^{*2}}$$

$$= 1 + \frac{j(j+1) + s(s+1) - l(l+1)}{2j(j+1)}$$

In an atom with several electrons, all the orbital angular momenta may be combined to give a single resultant $Lh/2\pi$, where L represents the vector sum of the l vectors for the individual electrons. The spin angular momenta may be combined similarly to give a single resultant $Sh/2\pi$. If the spin–orbital coupling is sufficiently strong, then these will combine in the same manner as outlined for l and s above to give a single resultant J. The magnetic moment of the whole atom is thus

$$g\beta\sqrt{J(J+1)}$$

APPENDIX ELEVEN

The Brillouin Function

Quantum-mechanical theory shows (Appendix 10) that the orientation of an atom to a magnetic field is restricted to certain values, so that the Langevin treatment of Section 12.4 is not strictly applicable. The total magnetic moment of an atom is $g\beta\sqrt{J(J+1)}$ and the component in the direction of the field is $g\beta m_J$ where m_J is a quantum number which can take the possible values J, $J-1, \ldots, -(J-1), -J$. The magnetic potential energy of an atom in an external field of flux density B is $-g\beta m_J B$. Then according to the Maxwell–Boltzmann statistics, the probability of an electron occupying such a state is proportional to

$$e^{g\beta m_J B/kT} = e^{m_J x}$$

where

$$x = g\beta B/kT$$

The total magnetic moment of N atoms of the system at a temperature T is

$$M = \frac{\sum_{m_J=-J}^{J} Ng\beta m_J e^{m_J x}}{\sum_{m_J=-J}^{J} e^{m_J x}}$$

$$= Ng\beta \frac{d}{dx}\left(\log \sum_{m_J=-J}^{J} e^{m_J x}\right)$$

$$= Ng\beta \frac{d}{dx}\left(\log \frac{e^{(J+1/2)x} - e^{-(J+1/2)x}}{e^{(1/2)x} - e^{-(1/2)x}}\right)$$

$$= NgJ\beta\left(\frac{2J+1}{2J}\coth\frac{2J+1}{2J}a - \frac{1}{2J}\coth\frac{a}{2J}\right)$$

$$= NgJ\beta B_J(a)$$

where

$$a = \frac{gJ\beta B}{kT}$$

305

and $B_J(a)$ is known as the *Brillouin function*. It can be shown that this function, as for the Langevin function, also gives the Curie law (equation (15.1)) for sufficiently low magnetic fields or high temperatures.

The Brillouin function is found to agree more closely to experimental results than the Langevin function and permits the appropriate values of g and J to be evaluated. In particular, for the iron group of transition elements, spin–orbital coupling does not hold and the experimental values of susceptibility can be accounted for by electron spin only.

Appendix Twelve

The Curie-Weiss Law

In a solid, the magnetic interaction between dipoles is not negligible and the effective field acting on a dipole must include the effect of the magnetization of the solid. Weiss allowed for this by considering an effective field

$$H_{\text{eff}} = H + \lambda M$$

where λ is the *Weiss field constant*. The magnetization is related to the effective field by Curie's law so that

$$M = \frac{C}{T}(H + \lambda M)$$

from which the susceptibility becomes

$$\chi = \frac{M}{\mu_0 H}$$

$$= \frac{C}{T(1 - \lambda C/T)}$$

$$= \frac{C}{T - \theta}$$

where

$$\theta = \lambda C$$

Appendix Thirteen

The Origin of Ferromagnetism

If, for temperatures below the Curie temperature, the effective field (Appendix 12) is substituted into the Brillouin function, then

$$\frac{M}{M_s} = B_J\left(\frac{gJ\beta\mu_0(H + \lambda M)}{kT}\right)$$

where

$$M_s = NgJ\beta$$

is the saturation magnetization when all the dipoles are aligned. If the external field is made equal to zero, then

$$\frac{M}{M_s} = B_J\left(\frac{gJ\beta\mu_0\lambda M}{kT}\right) \quad (A.29)$$

But also for $H = 0$

$$a = \frac{gJ\beta\mu_0\lambda M}{kT}$$

so that

$$\frac{M}{M_s} = \frac{kT}{gJ\beta\mu_0\lambda M_s}a$$

The Curie constant is, from equations (15.1) and (15.2), given by

$$C = \frac{\mu_0 M_s^2}{3kN} = \frac{\mu_0 gJ\beta M_s}{3k}$$

The Origin of Ferromagnetism

so that

$$\theta = \lambda C$$
$$= \frac{\mu_0 g J \beta \lambda M_s}{3k}$$

and therefore

$$\frac{M}{M_s} = \frac{T}{3\theta}a \qquad (A.30)$$

Equations (A.29) and (A.30) are two simultaneous equations for the reduced magnetism M/M_s as functions of a. These are shown qualitatively in Figure A.8. It is seen that when the slope of the straight line is less than $\frac{1}{3}$, i.e. $T < \theta$,

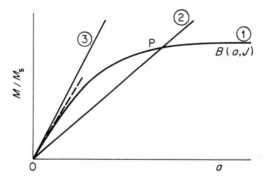

Figure A.8. Simultaneous equations for M/M_s as a function of a. (1) Brillouin curve (equation (A.29)), (2) straight line (equation (A.30)) for $T < \theta$, (3) straight line for $T > \theta$

then the curves intersect in two points. The intersection at the origin 0 corresponds to an unstable state, but as soon as a few atomic dipoles become lined up by chance, they will constitute a small magnetic domain and set up a field which will align more dipoles until a state represented by the other point of intersection P is attained, which represents a lower-energy stable state. If $T > \theta$, the slope is greater than $\frac{1}{3}$ and the curves intersect only at the origin showing that there can be no spontaneous magnetization above the Curie temperature.

From the coordinates of P we can derive a curve for M/M_s as a function of T/θ. Because the $B_J(a)$ functions will differ for systems with different J values, there will be a different curve for each value of J. Some examples are shown in

310 Properties of Materials for Electrical Engineers

Figure A.9. The experimental points for iron, cobalt and nickel would lie very close to the curve corresponding to $J = \frac{1}{2}$, which suggests that electron spin makes the only significant contribution to the magnetization.

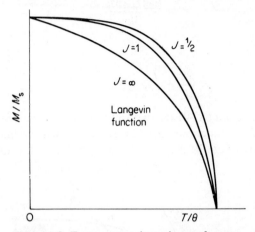

Figure A.9. Temperature dependence of saturation magnetization. The experimental points for iron, cobalt and nickel would agree most closely with the curve for $J = \frac{1}{2}$

Answers to Numerical Questions

Chapter 2

1. 1215×10^{-10} m; 18750×10^{-10} m.
2. 8.718×10^{-18} J.
3. 2.02×10^6 m s^{-1}; 11.6 eV.
4. 1.325×10^{-33} m.
5. (a) 4.355×10^{-23} kg m s^{-1};
   (b) 1.52×10^{-11} m;
   (c) 4.13×10^{11} m^{-1}.
6. 2.07×10^{-11} eV.
7. 5.64×10^5 m s^{-1}.

Chapter 3

1. 0.9706.
2. $I = \dfrac{N}{AL}\sqrt{\dfrac{2E}{m}}$.
3. 9.36×10^{-8} eV.
4. 2.27×10^{21}.
5. 5.29×10^{-11} m; 2.18×10^{-18} J.

Chapter 4

3. 1.55×10^{-10} m; 1.31×10^{-10} m; 14.4×10^{-10} m.

Chapter 5

5. 3.162 Å; 3.867 Å.

Chapter 6

1. 86·9.
2. 413 m s^{-1}; 3·02 × 10^{-10} m.
3. 1·079 N m^{-2}.
4. 1·342.
5. $\lambda \simeq 4$ m.
8. 37·55.

Chapter 7

2. 2·49 Å; 8910 kg m^{-3}.
3. 8806 kg m^{-3}; 72° 58'; 58° 30'.
4. 2·0393 Å; 1·4418 Å; 2·3548 Å.

Chapter 8

3. 269.
4. 150000.
5. 0·145; $2\frac{1}{2}$ per cent.

Chapter 9

2. 231·3 J kg^{-1} K^{-1}; 0·038 per cent.
3. 28000; 40.
4. 700 days.
5. 2·68 × 10^{-19} J atom^{-1}; 1·615 × 10^8 J kmol^{-1}.

Chapter 10

1. 5·04 × 10^{-19} J = 3·14 eV.
4. 1.
6. (c) 4·56 m.

Chapter 11

2. 7·0 m^2 V^{-1} s^{-1}; 7·0 m s^{-1}.
3. 415 J m^{-1} s^{-1} K^{-1}; 5·03 J m^{-1} s^{-1} K^{-1}.
4. (a) 4·37 eV; 26·6 × 10^4 A m^{-2} K^{-2}.
   (b) +0·227 eV.
5. 0·038 eV; 1·9 × 10^{-8} m from surface.
6. 3·8 eV.

Answers to Numerical Questions

Chapter 12

2. 0·47; 1·47.
3. $2 \cdot 0 \times 10^{-28}$ C m.

Chapter 13

1. $0 \cdot 35 \text{ m}^2 \text{ V}^{-1} \text{ s}^{-1}$.
2. $9 \cdot 45 \times 10^{-3}$ p.p.m.
3. $3 \cdot 41 \times 10^{-5}$ p.p.m.
4. $2 \times 10^{-4} \text{ m}^3 \text{ C}^{-1}$; $3 \cdot 7 \times 10^{22} \text{ m}^{-3}$.
5. $1 \cdot 25 \times 10^{-2}$ p.p.m.
7. $4 \cdot 5 \, \Omega\text{m}$; $0 \cdot 77 \, \Omega\text{m}$.
8. $2 \cdot 5 \times 10^{28} \text{ m}^{-3}$; $5 \cdot 32 \times 10^{-3} \text{ m}^2 \text{ V}^{-1} \text{ s}^{-1}$.

Chapter 14

1. $9 \cdot 29 \times 10^2 \, \Omega^{-1} \, \text{m}^{-1}$.
2. (a) $\Delta E = kT \log_e (N_D/n_i)$;
   (b) 0·73 V.
4. $6 \times 10^{17} \text{ m}^{-3}$.
5. $2 \cdot 5 \times 10^{23} \text{ m}^{-3}$; $2 \cdot 1 \times 10^9$ times.

Chapter 15

1. $2 \cdot 47 \times 10^{-10}$.
2. $2 \cdot 61 \times 10^{-6}$.
3. $1\frac{1}{12}$.
4. 2·18, 1·65, 0·60 Bohr magnetons.
5. $1 \cdot 07 \times 10^6 \text{ J m}^{-3}$.
6. $5 \cdot 7 \times 10^{-3}$ K; 628 K.

Bibliography

Anderson, J. C., *Magnetism and Magnetic Materials*, Chapman and Hall, London, 1968.
Azaroff, L. V. and Brophy, J. J., *Electronic Processes in Materials*, McGraw-Hill, New York, 1963.
Bacon, G. E., *Neutron Diffraction*, Oxford Clarendon Press, Oxford, 2nd ed., 1962, especially Chapters 6, 11 and 12.
Bates, D. R. (Ed.), *Quantum Theory*, Academic Press, New York, 1961.
Bates, L. F., *Modern Magnetism*, Cambridge University Press, Cambridge, 1951.
Brailsford, F., *Physical Principles of Magnetism*, Van Nostrand, London, 1966.
Cady, W. G., *Piezoelectricity*, McGraw-Hill, New York, 1946.
Craik, D. J., *Structure and Properties of Magnetic Materials*, Pion, London, 1972.
Evans, R. C., *Crystal Chemistry*, Cambridge University Press, Cambridge, 2nd ed., 1964.
Goldman, J. E. (Ed.), *The Science of Engineering Materials*, John Wiley, New York, 1957.
Greiner, R. A., *Semiconductor Devices and Applications*, McGraw-Hill, New York, 1961.
Hadfield, D. (Ed.), *Permanent Magnets and Magnetism*, Iliffe, London, 1962.
Hoselitz, K., *Ferromagnetic Properties of Metals and Alloys*, Oxford Clarendon Press, Oxford, 1952.
Jona, F. and Shirane, G., *Ferroelectric Crystals*, Pergamon, Oxford, 1962.
Lax, B. and Button, K. J., *Microwave Ferrites and Ferrimagnetics*, McGraw-Hill, New York, 1962.
Megaw, H. D., *Ferroelectricity in Crystals*, Methuen, London, 1957.
Mott, N. F., *Elements of Wave Mechanics*, Cambridge University Press, Cambridge, 1952.

Mott, N. F. and Sneddon, I. N., *Wave Mechanics and its Applications*, Oxford Clarendon Press, Oxford, 1948.

Pauling, L., *The Nature of the Chemical Bond*, Oxford Clarendon Press, Oxford, 3rd ed., 1960.

Roberts, J., H.F. *Applications of Ferrites*, English Universities Press, London, 1960.

Rose, R. M., Shepard, L. A. and Wulff, J., *The Structure and Properties of Materials, Volume IV, Electronic Properties*, John Wiley, New York, 1966.

Schmidt, A. X. and Marlies, C. A., *Principles of High-Polymer Theory and Practice*, McGraw-Hill, New York, 1948.

Seitz, F., *Modern Theory of Solids*, McGraw-Hill, New York, 1940.

Shockley, W., *Electrons and Holes on Semiconductors*, Van Nostrand, New York, 1950.

Slashkov, Y. M., *The Metallurgy of Semiconductors* (translated by Bradley, J. E. S.), Pitman, London, 1961.

Smit, J. and Wijn, H. P. J., *Ferrites*, Phillips Technical Library, Eindhoven, 1959.

Troup, G., *Masers and Lasers*, Methuen, London, 2nd ed., 1963.

Waldron, R. A., *Ferrites*, Van Nostrand, London, 1961.

Warschauer, D. M., *Semiconductors and Transistors*, McGraw-Hill, New York, 1959.

Whitehead, S., *Dielectric Breakdown of Solids*, Oxford Clarendon Press, Oxford, 1951.

Index

Absorption band 182
Acceptor 203
Acid 122
Actinide 8
Activation energy 135, 183
Addition polymerization 121
  polymer 122
Alcohol 122
Alcomax 263
Aliphatic compound 120
Alkali
  halide 251
  metals 254
    band structure of 141
Alkaline earth metals 254
  band structure of 142
Allowed energy band 140
Alnico 263
α particle 12, 38
Alumina 114, 185
Aluminium 215, 216
  antimonide, properties of 231
  arsenide, properties of 231
  phosphide 217
    properties of 231
Amino-acid 123
Ammonia molecule, shape of 73
Amorphous solid 95
Ampere, definition 275
Ampere's law 249
Amplitude-modulated wave 19
Andrews, experiments on CO_2 88
Ångström, definition 276
Anisotropy
  energy 258
  magnetic 257
Antiferromagnetic 247, 265
  material, susceptibility of 266
Antimony 201, 215

Araldite 126
Aromatic compound 120
  hydrocarbon 120
Arrhenius' rate law 137
Arsenic 215
Atom 1
Atomic
  heat 132
  mass unit 1
  number 2, 6
  volume 6
  weight 1, 2
Avogadro's hypothesis 80
  number 81, 277

Band structure 133
  of alkali metals 141
  of alkaline earth metals 142
  of solids 140
Bardeen 220
Barium
  carbonate 166
  titanate 192
Base 234
Benzene 120
Binding energy 71
Black-body radiation 9
Body-centred cubic structure 104
Bohr
  magneton 249, 257, 277
  radius 203
    Rutherford atom 12
Boltzmann's constant 81, 136, 277
  distribution 134, 136
Bonding
  covalent 64, 67, 70
  interatomic 63
  ionic 69, 70
  metallic 76

Bonding (*cont.*)
  polar   69
  secondary   117
  Van der Waals   117
Boric oxide   116
Boyle's law   79
Breakdown voltage   179
Brillouin
  function   305
  zone   151, 197
Brittle   114
Butane   118
Butene   120
Butylene   120

Calcium titanate   194
Candela, definition   275
Carbon tetrachloride   118
Cadmium oxide, properties of   231
Cadmium selenide, properties of   231
Cadmium sulphide   217
  photocell   232
    response of   232
  properties of   231
Cadmium telluride, properties of   231
Caesium chloride, structure   109
Cellulose fillers   186
Celsius degree, definition   276
Centre of oscillation   95
Ceramics   111, 185, 186
Charles' law   79
Chloroform   118
Chloroprene   126
Chromium   265
Clausius–Mosotti equation   178
Close-packed hexagonal structure   106
Cobalt   255, 257, 265
  magnetization curves for   258
Coefficient of diffusion   87
Coercive
  field   190
  force   262
Collector   234
Compton effect   15
Concentric shells   59
Condensation polymer   123
  polymerization   122
Conduction band   196
Conductivity, thermal   163
Conductor   155
Copolymer   123
Copper   215, 254
Covalent bond   59, 64, 70
  directional properties of   71
Covalent bonding   185
Coulomb, definition   276

Cracking   114
Critical
  pressure   88
  temperature   88
  volume   88
Crystal
  structure   95
  systems   101
Cubic
  system   97, 102
  structure, body-centred   104
    face-centred   102
Curie
  law   253, 306, 307
  temperature   191, 255
Curie–Weiss law   253, 266, 307
Czochralski method   218

Davisson   18
de Broglie's hypothesis   18
Debye   133
  temperature   133, 144, 163
Degenerate   46, 58, 73
Degree Celsius, definition   276
Degree of polymerization   121
Density of states in semiconductor   197
Depletion layer   236
Diamagnetic   247, 250
  susceptibility   251
Diameter of gas molecule   84
Diamond   195, 215
  structure   107
Dielectric
  constant   174
  strength   184
Dielectrics   173
  conductivity of   183
Diffraction of electrons   18
Diffusion   212
  coefficient of   87
  in solid   137
  of gases   79, 84, 87
Diffusion rate   183
Dipolar polarization   174
Dipole
  electric   71, 173
  magnetic   249
  moment, permanent   77
Dirac   58
Discrete localized state   203
Domain
  electric   189
  magnetic   255, 259
  wall   259
Donor   201
  level   203

Index

Doping 201
Doublet structure 58
Drain 236
Dulong and Petit's law 132

Eddy current loss 263, 273
Effective
  density of states 199
  mass 160, 198
Eigenfunction 29
Eigenvalue 29, 42
Einstein 15, 18
  relation 183, 212
Elastomer 125
Electric dipole 71, 173
Electrical
  breakdown 184
  conductivity 79, 115, 158
  susceptibility 175
Electron
  diffraction 18
  in hydrogen atom 47
  mass of 277
  orbital 71
  spin 57, 253
  spins, interaction between 256
  structure of the elements 59
  volt 12, 276, 277
Electron-hold
  pair generation 219
  product 223
Electronegative element 59
Electronic charge
  value of 277
Electronic polarization 174, 182
Electropositive element 59
Electrostriction 188
Element 1
Elements, electron structure of 59
Emitter 234
Energy
  band 140, 195
  gap 173
  levels, splitting of 58
Epoxy
  group 126
  resin 126
Equipartition of energy 129
Ethane 118
  bonding in 74
Ethyl 122
Ethylene 119
  chloride 120
Exchange
  energy 255, 259
  forces 255

Exothermic reaction 118

Face-centred cubic structure 102
Farad, definition 276
Faraday rotation 272
Fermi
  distribution curve 143
  function 199
  level 143, 153, 166, 199, 222, 253
    position of 205
    variation with temperature 205
Fermi–Dirac
  distribution 134, 298
  statistics 143, 255, 295
Ferric oxide 217
Ferrimagnetic 111, 247, 267
  materials 268
Ferrite 268
  toroid 273
Ferrites
  applications of 273
  electromagnetic wave propagation in 271
Ferroelectric 189
Ferromagnetic 247, 254
  materials 155, 261
    applications of 263
Ferrous oxide 257
Ferroxdure 271
FET 235
Fick's laws 138
Field
  effect transistor 235
  emission 167
Fillers 127, 186
Flow temperature 125
Forbidden
  energy band 140
  gap 148
Formaldehyde 124

Gadolinium 257
Gallium 203, 215
  antimonide, properties of 231
  arsenide 217, 219, 229, 233
    properties of 231
  phosphide, properties of 231
γ-ray 17
  microscope 22
Garnet 273
Gas
  constant 277
  diffusion of 79, 84, 87
  kinetic theory of 79, 80
  thermal conductivity of 79, 84, 86
  viscosity of 79, 84, 86

Gaseous diffusion 87
Gate 236
Germer 18
Germanium 196, 215, 229, 233
  properties of 231
Glass 112, 185, 186
Glass-ceramics 117
Glass-transition temperature 125
Glassy solid 95
Gold 254
Graded junction 226
Graphite 114
Gray tin 196, 215

Hall
  constant 210
  effect 209
    device 237
Hartree 57
Heat of reaction 135
Heisenberg 255
Heisenberg's uncertainty principle 21
Henry, definition 276
Hertz, definition 276
Hexagonal close-packed structure 106
  system 97, 102
Hole 196
Homologous series 118
Hunt's rule 250
Hybrid orbital 73
Hydrocarbon
  aromatic 120
  saturated 117
  unsaturated 119
Hydrogen
  bond 77
  bridge 77
  molecule 64
Hysteresis
  loop 262
  loss 263, 273
  magnetic 262

Impurity concentration 205
Indices
  of directions, Miller 100
  of planes, Miller 98
Indium 215
  antimonide 217, 238
    properties of 231
  arsenide 217, 238
    properties of 231
  phosphide, properties of 231
  sulphide 232
Inert gases 251
Inorganic compounds 111
Insulating materials 173

Insulator 111, 156
Interatomic bonding 63
Interfacial polarization 174, 186
Intermolecular forces 76
Intrinsic semiconductor 199
Inverse piezoelectric effect 187
Inversion layer 220
Inverted spinel structure 268
Ionic
  bonding 69, 70
  crystals 108
  polarization 174, 182
Ionization 79
  energy 203
Iron 250, 255, 257, 265
  magnetization curves for 258
Isomer 118
Isoprene 125
Isotope 1

Joule, definition 276
Junction rectifier 229

Kaolin 111, 114
Kelvin, definition 275
Kelvin relations 170
Kilogramme
  definition 275
  mole 81
    definition 276
Kinetic energy of gas molecule 129
  theory of gases 79, 80
kmol 81
k-space 46, 150

Landé splitting factor 271, 304
Langevin function 179
Lanthanide 8
Larmor
  frequency 271
  precession 251, 271
Lattice
  constant 2
  point 95
Law
  Ampere's 249
  Arrhenius' rate 137
  Boyle's 79
  Charles' 79
  Curie 253, 306, 307
  Curie–Weiss 253, 307
  Dulong and Petit's 132
  Fick's 138
  Lenz's 251
  Maxwell's 9
  Moseley's 61

Index

Law (*cont.*)
  Newton's 9
  of periodicity 4
  Stefan's 10
  Wein's 9
  Wiedmann–Franz 164
Lead
  selenide, properties of 231
  telluride, properties of 231
Lenz's law 251
Light, velocity of 277
Litre, definition 276
Loss angle 183

Magnesia 111, 114, 185, 217
Magnetic
  anisotropy 257
  behaviour 60
  dipole 249
  domain 255
  garnets 270
  hysteresis 262
  materials 246
  permeability 246
  susceptibility 246
Magnetite 269
Magnetogyric ratio 249, 271, 303
Magnetoplumbite structure 268, 270
Magnetostatic energy 259
Magnetostriction 258
Magnetostrictive energy 258
Manganese 257, 265
  copper ferrite 270
  magnesium ferrite 270, 274
  zinc ferrite 269
Manganous oxide 266
Many-electron atom 56
Mass, effective 160
Mattheison's rule 163
Maxwell 129
  distribution law 9, 81
Maxwell–Boltzmann
  distribution 134, 211, 293
  statistics 295, 305
Mean free path 82, 88
Mendeleev 1, 5
Metal-oxide semiconductor transistor 236
Metal-semiconductor rectifier 221
Metallic bond 76
Metastable state 135
Methane 118
  molecule, shape of 73
Methyl 118, 122
  chloride 118
  methacrylate 123

Methylene chloride 118
Metre, definition 275
Mica 185
Microminiature solid state circuits 240
Miller
  index notation 98
  indices 98
Miller–Bravais system 100
Minority carrier lifetime 219
Mobility 207
  of electron 159, 161
Mole 81
Molecular
  spectra 68
  weight 1
Molecule 1
Monoclinic system 97, 102
Monomer 121
Moseley's law 61
MOST 235

Natural rubber 125
Negative ion 69
Neoprene 126
Neutron 14
  mass of 277
Newton, definition 276
Newton's laws 9
Nickel 255, 257, 265
  magnetization curves for 258
  oxide 217
  zinc ferrite 269, 273, 274
Nitrogen 216
Non-polar molecules 124
Normalization of wave function 30
n-type semiconductor 166
Nucleus 12
Nylon 123

Ohm, definition 276
Ohmic contact 225
Olefine 119
Orbital angular momentum 250
Organic
  compounds 111
  semiconductor 217
Orthorhombic system 97, 102
Oscillator, simple harmonic 52
Overlapping wave functions 63

Paraffin 118
  wax 118
Paramagnetic 247, 252
  susceptibility 252
Pauli's exclusion principle 57, 68, 133, 134,
    136, 256, 296

Peltier
  coefficient 170
  effect 170
Pentane 118
Perfect gas 79
Periodic
  table 1, 4
  field effect of 146
Permalloy 264
Permanent magnet 262, 263
Permeability
  absolute 246
  of space 277
  relative 246
Permittivity 174
  of space 277
  relative 175
Perspex 123
Petit's, Dulong and, law 132
Phenol 121, 124
  formaldehyde 124
    resins 186
Phonon 163
Phosphorus 215
  pentoxide 116
Photocell, semiconductor 230
Photoelectric
  effect 11
  work function 12, 165
Photon 11
Photovoltaic effect 232
π bond 74, 120
Piezoelectric 111, 186
Planck's constant 11, 277
  radiation formula 11
Planes of a form 100
Plasticizers 128, 186
p–n junction 226
Point-contact germanium rectifier 225
Polar bonding 69
  materials 174
  molecules 124
Polarizability 176
Polarization 173, 182
  dipolar 174
  electronic 173
  interfacial 174
  ionic 173
Polyamide 123
Polyester 123
Polyethylene 121
Polymer 121
Polymerization 121
  addition 121
  condensation 121
  degree of 121

Polymethylmethacrylate 123
Polypropylene 122
Polystyrene 122, 186
Polytetrafluoroethylene 185, 186
Polythene 121, 185, 186
Polyvinylchloride 122, 184
Positive ion 69
Potential barrier
  rectangular 36
  variable width 37
Potential jump 32
Potential well
  deep 40
  low sided 42
  three-dimensional 45
Power factor 183
Primitive cell 96
Propane 118
Propene 119
Propylene 119, 122
Proton 14
  mass of 277
p.t.f.e. 185
p.v.c. 122, 185, 186
Pyroelectric 188
  materials 108

Quantum 11
  mechanical tunnelling 167
  number 42, 49, 53, 68, 250
Quartz 185, 188

Radial charge density 51, 59
Radioactive emission 38
  isotope 137
Rare earth 8, 60, 254
  iron garnets 268
Recombination centre 219
Rectangular potential barrier, effect of 36
Rectifier
  junction 229
  metal-semiconductor 221
  point contact 225
Rectifying contact 222
Refractive index 178
Refractory 114
Relative
  permeability 246
  permittivity 175
Relativistic mass 15
Remnance 262
Remnant polarization 190
Resistivity 159
  temperature coefficient of 163
Resonance absorption 182

Index

Rest
  mass 15
  energy 15
Rhombohedral system 97, 102
Richardson–Dushmann equation 165
  proof of 298
Rochelle salt 188, 191
Rotational
  energy 68
  mode 68
Rubber 186
  natural 125
Rutherford atom, Bohr- 12
Rutile 185, 194
Rydberg constant 14, 277
Rydberg, definition 276

Saturated hydrocarbon 117
Saturation
  magnetization 255
  vapour pressure 92
Schottky effect 166
Schrödinger's equation 33, 40
  conditions for solution 29
  formulation of 28
  solution for hydrogen atom 278
  solution for simple harmonic oscillator 285
Second, definition 275
Secondary bonds 117
Seebeck potential 167, 222
Seed crystal 218
Segregation coefficient 218
Selenium 216
  properties of 231
Semiconductor 111, 156, 195
  density of states in 197
  extrinsic 203
  intrinsic 199
  materials, purification of 218
  n-type 203
  organic 217
  p-type 203
σ bond 74, 120
Silica 114
Silicates 111
Silicon 196, 215, 233
  properties of 231
Silicon carbide 111, 115, 216, 229, 240
  properties of 231
Silicon iron 264, 273
Silver 254
Simple harmonic oscillator 52
Sodium chloride, structure 109
Softening temperature 116
Solar battery 232
Solid-state diffusion 137

Source 236
Space lattice 95
Specific heat 6
  capacity of a gas 131
  capacity of a solid 131
Spectra, molecular 68
Spherical symmetry of full p and d levels 288
Spin angular momentum 250
Spin–orbital coupling 302
Spinel 268
Spontaneous magnetization 255
Stable state 135
Stationary state 40
Statistical thermodynamics 129
Stefan's law 10
Stefan–Boltzmann constant 10, 277
Step junction 226
Strontium
  carbonate 166
  titanate 194
Structure cell 96
Styrene 122
Superconducting material 247, 252
Superexchange interaction 265
Surface
  energy 92
  state 220
  tension 91
Susceptibility
  diamagnetic 246
  electrical 175
  magnetic 246
  paramagnetic 252

Tellurium 216
  properties of 231
Temperature coefficient of resistivity 163
Tesla, definition 276
Tetrafluoroethylene 123
Tetragonal system 97, 102
Thermal conductivity 163
  of gases 79, 84, 86, 114
  of liquid 93
Thermal
  equilibrium 134
  expansion 133
    differential 114
  insulator 114
Thermionic emission 156, 224
Thermistor 238
  characteristics of 239
Thermocouples 167
Thermoelectric power 170
Thermoelectricity 167
Thermoplastic resin 125
Thermosetting resin 125, 185, 186

Thomson, G.P. 18
Thomson
  coefficient 170
  effect 169
Thorium 166
  dioxide 166
Threshold frequency 12
  voltage 237
Time-delay switching 240
Tin 196
Toluene 121
Transducer 188
Transistor
  field effect 235
  junction 235
  metal-oxide semiconductor 236
Transition elements 6, 8, 60
  metals 254
Transmission coefficient 37
Transparency 37
Transport
  phenomena 84
  properties 84
Triclinic system 97, 102
Tungsten 166
Tunnel diode 232
  characteristics of 234
Tunnel effect 37
Tunnelling, quantum mechanical 167

Uncertainty principle, Heisenberg's 21
Unified atomic mass unit 1, 277
Unit cell 96
Unsaturated hydrocarbons 119
Urea 124
  formaldehyde 124

Valence band 196
Valency 6, 108
Van der Waals
  bonding 185
  bonds 117
  equation 89
  forces 77, 91, 121, 124
Van Vleck 267
Vapour pressure 92
Varistor 240
Velocity of light 277
Vibrational mode 68

Vinyl chloride 122
Viscoelastic strain 124
Viscosity
  of gas 79, 84, 86
  of liquid 92
Vitreous structure 115
Volt, definition 276
Voltage-variable capacitance diode 234
Vulcanizing 126

Water molecule, shape of 72
Watt, definition 276
Wave
  amplitude modulated 19
  equation, simple 24
  function 19, 42, 49, 53
  functions, overlapping 63
  intensity 26
  mechanics 9
  number 19, 25
  vector 198
Weber, definition 276
Wein's displacement law 9
Weiss 253, 255
  field constant 307
  molecular field 255
Wiedmann–Franz law 164
Work function 164, 166
  photoelectric 12, 165
Wurtzite 189
  structure 109

X-ray 17, 60
  diffraction 98
  spectra 60
Xylol 121

Zeeman effect 302
Zener diode 230
  characteristics of 230
Zinc 215
  oxide 217
    properties of 231
  sulphide 189
Zincblende 189
  structure 109
Zone
  levelling 218
  refining 218